EPIGENETIC RISKS OF CLONING

EPIGENETIC RISKS OF CLONING

EDITED BY

Akio Inui

Kagoshima University
Kagoshima, Japan

Taylor & Francis
Taylor & Francis Group

Boca Raton London New York Singapore

A CRC title, part of the Taylor & Francis imprint, a member of the Taylor & Francis Group, the academic division of T&F Informa plc.

Published in 2006 by
CRC Press
Taylor & Francis Group
6000 Broken Sound Parkway NW, Suite 300
Boca Raton, FL 33487-2742

Printed in the United States of America on acid-free paper
10 9 8 7 6 5 4 3 2 1

International Standard Book Number-10: 0-8493-2527-7 (Hardcover)
International Standard Book Number-13: 978-0-8493-2527-4 (Hardcover)
Library of Congress Card Number 2005048532

Library of Congress Cataloging-in-Publication Data

Epigenetic risks of cloning / edited by Akio Inui
p. cm.
Includes bibliographical references and index.
ISBN 0-8493-2527-7 (alk. Paper)
1. Cloning. 2. Epigenesis. I. Inui, Akio, 1950-

QH442.2.E65 2005
660.6'5—dc22 2005048532

T&F informa

Taylor & Francis Group
is the Academic Division of T&F Informa plc.

Visit the Taylor & Francis Web site at
http://www.taylorandfrancis.com

and the CRC Press Web site at
http://www.crcpress.com

Preface

Cloning technology has the potential to be a valuable tool not only in basic research, but also in agriculture and clinical medicine. The agricultural and clinical applications that are being explored include reproductive cloning of farm animals and therapeutic cloning for human cell, tissue, and organ replacement [1,2]. Embryos produced by nuclear transfer from a patient's somatic cell offer one potential source of embryonic stem (ES) cells [3,4] that would be histocompatible with the patient's cells and in principle be a source of any other cell type. Studies in animal models show that transplantation of ES-derived cells can successfully treat a variety of chronic diseases, including cardiovascular diseases, diabetes, and traumatic spinal cord injury, that underlie the promising role of human ES cells in tissue regeneration and modern medicine [5].

Several laboratories have used a variety of somatic cell types to create cloned animals, including sheep, cattle, mice, pigs, and goats; the list is ever expanding. The present procedures have proved to be repeatable, but are very inefficient when only between 1 and 4% of reconstructed embryos typically develop to adulthood [1,2]. The low overall success rate is the cumulative result of inefficiencies at all stages of development, although species and donor cell types may differ in the precise pattern of loss. In addition to embryonic loss, somatic cell nuclear transfer is also associated with very high rates of fetal, perinatal, and neonatal loss, and production of abnormal offspring [1,2,6]. Common abnormalities include respiratory distress, increased birth weight, and major cardiovascular abnormalities. Other abnormalities that may develop later include failure of the immune system; structural abnormalities of the brain and other viscera; accelerated aging; and obesity, which may be influenced by species, genetic background, or donor cell type.

Increasing evidence from a range of mammals shows a propensity for epigenetic errors with embryo technologies; if paralleled in human embryos, the effect on tumorigenic and differentiation properties of ES cells needs to be established [7]. Epigenetic risks are also related to assisted reproductive technologies in humans as known for imprinting disorders such as Beck–Wiedemann and Angelman syndromes. Therefore, it is important to evaluate the consequences of cloning in resulting embryo and offspring before widespread use of the technology.

We hope that this book helps the reader understand embryonal, fetal, perinatal, neonatal, and postnatal development of clones of various species for further technological advances.

REFERENCES

1. Wilmut, I., and Paterson, L. Somatic cell nuclear transfer, *Oncol. Res.* 13, 303, 2003.
2. Wilmut, I. et al. Somatic cell nuclear transfer, *Nature* 419, 583, 2002.
3. Cowan, C.A. et al. Derivation of embryonic stem-cell lines from human blastocysts, *N. Engl. J. Med.* 350, 1353, 2004.
4. Hwang, W.S. et al. Evidence of a pluripotent human embryonic stem cell line derived from a cloned blastocyst, *Science* 303, 1669, 2004.
5. Stojkovic, M. et al. Derivation, growth and applications of human embryonic stem cells, *Reproduction* 128, 259, 2004.
6. Tamashiro, K.L. et al. Phenotype of cloned mice: development, behavior, and physiology, *Exp. Biol. Med. (Maywood)* 228, 1193, 2003.
7. Allegrucci, C. et al. Stem-cell consequences of embryo epigenetic defects, *Lancet* 364, 206, 2004.

Editor

Akio Inui completed his M.D. and Ph.D. degrees at Kobe University. In 1984, he was appointed assistant professor at the same university and became associate professor in 1998. Between 1985 and 1989, he also served as chief physician of the Medical Ward at Kobe University Hospital. Since 2001, Dr. Inui has been engaged as associate professor in the Division of Diabetes, Digestive and Kidney Diseases in the Department of Clinical Molecular Medicine at Kobe University Graduate School of Medicine. He is recognized for his studies in elucidating the role of gut–brain peptides in the regulation of gastrointestinal motility, feeding, and body weight, and the identification of the ghrelin–neuropeptide Y pathway from stomach to brain in the pathogenesis of obesity and cachexia. He has an interest in the role of gut–brain peptides in the development of behavior in animal models, including clones.

Dr. Inui received the Janssen Award of the American Gastrointestinal Association in 2004. He is editor of *Peptides*, *Nutrition*, and the *International Journal of Oncology*. Between 2000 and 2002, he also edited the *International Journal of Molecular Medicine*. His efforts are now focused on translational research on peptides that bridges the gap between basic and clinical disciplines for better understanding and management of human behavioral disorders, including obesity, cancer cachexia, and eating disorders.

Contributors

Pierre G. Adenot
INRA
Unite de Biologie du Developpement et Reproduction
Jouy-en-Josas, France

Laurent Boulanger
INRA
Unite de Biologie du Developpement et Reproduction
Jouy-en-Josas, France

Thomas D. Bunch
Animal Dairy and Veterinary Sciences Department, Center for Integrated Biosystems
Utah State University
Logan, Utah

Mireille Challah–Jacques
VIVALIS
Site Bio Quest
Saint-Herblain, France

Aimee S. Chang
Department of Obstetrics and Gynecology
Washington University School of Medicine
St. Louis, Missouri

Patrick Chesné
INRA
Unite de Biologie du Developpement et Reproduction
Jouy-en-Josas, France

Peter Chrenek
RIAP
Nitra, Slovak Republic

Nathalie Daniel
BioProtein Technologies
Paris, France

Michael R. DeBaun
Department of Pediatrics, Division of Genetics
Washington University School of Medicine
St. Louis, Missouri

Robert Feil
Institute of Molecular Genetics, CNRS
Montpellier, France

Martha C. Gómez
Audubon Center for Research of Endangered Species and Department of Animal Sciences
Louisiana State University
New Orleans, Louisiana

J.B. Gurdon
Wellcome Trust/Cancer Research
U.K. Gurdon Institute and Department of Zoology
Cambridge, United Kingdom

M. Hirabayashi
National Institute for Physiological Sciences
Okazaki, Aichi, Japan

S. Hochi
Faculty of Textile Science and Technology
Shinshu University
Ueda Nagano, Japan

Yoko Kato
Laboratory of Animal Reproduction, College of Agriculture
Kinki University
Nakamachi, Nara, Japan

Guanpeng Li
Animal Dairy and Veterinary Sciences Department, Center for Integrated Biosystems
Utah State University
Logan, Utah

Pasqualino Loi
Department of Comparative Clinical Services
University of Teramo
Teramo, Italy

Kelle H. Moley
Department of Obstetrics and Gynecology
Washington University School of Medicine
St. Louis, Missouri

Akira Onishi
Embryonic Technology Laboratory, Developmental Biology Department
Institute of Insect and Animal Sciences, National Institute of Agrobiological Sciences
Tsukuba, Ibaraki, Japan

C. Earle Pope
Audubon Center for Research of Endangered Species
Louisiana State University
New Orleans, Louisiana

Jean-Paul Renard
INRA
Unite de Biologie du Developpement et Reproduction
Jouy-en-Josas, France

Randall R. Sakai
Department of Psychiatry
University of Cincinnati College of Medicine
Cincinnati, Ohio

Kellie L.K. Tamashiro
Neuroscience Program and Department of Psychiatry
University of Cincinnati College of Medicine
Cincinnati, Ohio

Yukio Tsunoda
Laboratory of Animal Reproduction, College of Agriculture
Kinki University
Nakamachi, Nara, Japan

Dirk K. Vanderwall
Northwest Equine Reproduction Laboratory, Department of Animal and Veterinary Science
University of Idaho
Moscow, Idaho

Xavier Vignon
INRA
Unite de Biologie du Developpement et Reproduction
Jouy-en-Josas, France

Teruhiko Wakayama
Institute for Biogenesis Research
University of Hawaii School of Medicine
Honolulu, Hawaii
Laboratory for Genomic Reprogramming, Center for Developmental Biology
Kobe, Japan

Michael Wangler
Department of Pediatrics, Division of Genetics
Washington University School of Medicine
St. Louis, Missouri

Kenneth L. White
Animal Dairy and Veterinary Sciences Department, Center for Integrated Biosystems
Utah State University
Logan, Utah

Gordon L. Woods
Northwest Equine Reproduction Laboratory, Department of Animal and Veterinary Science
University of Idaho
Moscow, Idaho

Yukiko Yamazaki
Institute for Biogenesis Research
University of Hawaii School of Medicine
Honolulu, Hawaii

Ryuzo Yanagimachi
Institute for Biogenesis Research
University of Hawaii School of Medicine
Honolulu, Hawaii

Table of Contents

1 Health Consequences of Cloning Mice

Kellie L.K. Tamashiro, Randall R. Sakai, Yukiko Yamazaki, Ryuzo Yanagimachi, and Teruhiko Wakayama

CONTENTS

ABSTRACT

To date, cloning from adult somatic cells has been successful in at least 10 species. Although generating viable cloned mammals from adult cells is technically feasible and this list will only continue to grow with time, prenatal and perinatal mortality is high and live cloned offspring have not been without health problems. The success of many of the proposed applications of the cloning technique obviously depends upon the health and survival of founder animals generated by nuclear transfer. This review summarizes the health consequences of cloning in mice and discusses possible mechanisms through which these conditions may arise. These studies have further significance because other assisted reproductive techniques (ART) also involve some of the same procedures used in cloning; some reports have indicated that offspring generated by ART display aberrant phenotypes as well. At the moment, the long-term consequences of mammalian cloning remain poorly characterized. Data available thus far suggest limiting use of this technology until numerous questions are addressed and answered.

1.1 INTRODUCTION

It was once believed that cloning mammals from differentiated adult somatic cells was impossible [1], but it is now known that this is not the case. In 1997, Ian Wilmut and colleagues in Scotland reported that they had successfully generated a cloned mammal, a sheep named Dolly, using nuclear transfer of nuclei from an adult mammary gland cell [2]. Initially, that Dolly was cloned from an adult somatic cell was in doubt because she was the only cloned sheep produced and efforts to replicate that feat failed. A year later, two independent groups, working with two different species, cast away that doubt, thus confirming that mammalian cloning using an adult somatic cell was indeed possible [3,4].

Since those first reports, other species have been cloned from adult cells, including the pig [5]; goat [6]; cat [7]; rabbit [8]; mule [9]; horse [10]; and rat [11]. The possible applications are numerous and range from production of genetically modified livestock for human consumption to generation of replacement tissue for transplant therapy, and even human reproduction. Although the efficiency of cloning in all species is extremely low [12] and cloned animals exhibit various health problems, cloning technology will improve as scientists are allowed to determine the optimal conditions for cloning. As with all new technologies, it is important to evaluate the potential short- and long-term consequences prior to widespread use.

This review summarizes studies conducted in mice that begin to document the consequences of mammalian cloning from adult somatic cells. Selecting the mouse as the animal model offers the advantages of its easy handling and well-characterized behavior and genetics, as well as a short lifespan that permits longitudinal studies of these animals in a relatively short time span. Although the phenotype of cloned offspring has been published, it is important to note that in some cases the results have been mixed, suggesting that the consequence of cloning is highly unpredictable.

Additionally, although some altered phenotypes in cloned animals may be attributable to the cloning process, additional influences may arise from interlaboratory differences in technical proficiency. Indeed, some studies report phenotypic variations in animals generated from normal as well as reconstructed embryos exposed to *in vitro* culture conditions [13–17]. Thus, it is critical that the appropriate control groups are included in order to differentiate clone-specific phenotypes from those resulting from technical aspects of the technique [18,19].

Of equal, or perhaps greater, significance is the fact that many of the technical manipulations involved in cloning are also used in assisted reproductive techniques (ART) used in humans, including *in vitro* fertilization (IVF); intracytoplasmic sperm injection (ICSI); exchange of cytoplasm between oocytes; and *in vitro* culture. These techniques have been used extensively in humans for a short period of time. The oldest IVF baby is currently in her mid 20s; however, the long-term consequences of ARTs have not been determined. Perturbations in development resulting from *in vitro* manipulation of embryos have been documented in sheep, cattle, and mice [13,20], suggesting that these techniques are not innocuous to offspring. Systematic longitudinal studies must be done using cloning and ARTs, and care must be exercised in dissecting the effects of these techniques by including control groups to determine the contribution of such manipulations to any aberrations in offspring.

1.2 METHODS

Female B6C3F1 cloned mice were generated by the "Honolulu technique" to study the behavioral and physiological consequences of cloning on offspring [3]. Briefly, female mouse clones were generated by microinjection of cumulus cell nuclei from adult (8 to10 weeks old) B6C3F1 (C57BL/6 × C3H/He) hybrid mice into enucleated oocytes collected from adult (8 to 10 weeks old) B6D2F1 (C57BL/6 × DBA/2) mice. Preimplantation embryos were transferred into pseudopregnant CD-1 surrogate mothers. Pups were delivered at 19.5 days postcoitum (d.p.c.) by cesarean section and placed with the litters of lactating CD-1 foster mothers to be raised.

In addition, two control groups were employed. Members of the first group were age- and strain-matched mice generated by natural mating ("STOCK" controls). The second control group ("IVEM" controls) also consisted of age- and strain-matched mice. These animals were generated to control for several of the *in vitro* manipulations and procedures to which the cloned embryos were subjected, including *in vitro* culture until the two-cell stage and embryo transfer into surrogate mothers. IVEM mice were delivered by cesarean section at 19.5 d.p.c. and the pups were cross-fostered to lactating CD-1 foster mothers.

1.3 CLONING THE MOUSE

"Cumulina" was the first mouse to be cloned from an adult somatic cell, a cumulus cell [3]. Since this first report, other researchers have produced cloned mice from embryonic stem (ES) cells [21–24]; fetal neurons [25]; fibroblasts [26,27]; and immature Sertoli cells [28]. It is important to point out that the cloned mice produced in these studies were generated by piezo-driven microinjection of isolated cell nuclei. This is in contrast to other studies, including those examining phenotypes of cloned cattle, which used whole cell injection and electrofusion to generate cloned embryos. The two protocols have inherent differences, including the degree of mechanical trauma to the oocyte and reconstructed embryo and the amount of cytoplasmic contamination. Although the differences between these two protocols have been addressed [27], they have not yet been investigated thoroughly, particularly in regard to the potential long-term consequences on the phenotype of offspring.

The method of activation of the resulting embryo differs between the Honolulu technique and electrofusion method. The Honolulu technique relies on chemical activation by Sr^{2+} and a delay between nuclear transfer and activation; the electrofusion method involves simultaneous activation by the electric pulse during fusion of the donor cell with the enucleated oocyte. It has been demonstrated that the method of activation does not have a significant effect on postimplantation embryo development or the birth of live offspring [29]; however, whether it produces differences in the phenotype of offspring is not known.

1.4 EMBRYONIC AND PRENATAL DEVELOPMENT

As mentioned earlier, the success rate (percentage of reconstructed embryos that develop to full term) of cloned mice is very low and currently remains at about 1% for cumulus cells and 2 to 3% for ES cells. However, differences between the use of these two cell types emerge when progress to successive developmental stages is examined in detail (Figure 1.1) [21,30]. With cumulus cells, 55% of cloned embryos developed *in vitro* to the morula/blastocyst stage. Evaluation of *in utero* development beyond the morula/blastocyst stage was made at the time of cesarean section, which took place 16 days after embryo transfer; 35% of cloned embryos appeared to have implanted and 1% developed to live pups, with no dead fetuses.

In contrast, only 30% of cloned embryos completed preimplantation development when ES cells were used as nuclear donors. *In utero*, 20% showed evidence of implantation, 4% developed into a fetus, and 2% were alive at term. Many placentae (without fetuses) and dead fetuses arrested at 15 to 17 d.p.c. were observed on cesarean section. It is suggested that the specific restriction points within a successful cloning process may vary with the source of cells. The mouse is a preferred organism to study the parameters governing the cloning phenomenon systematically.

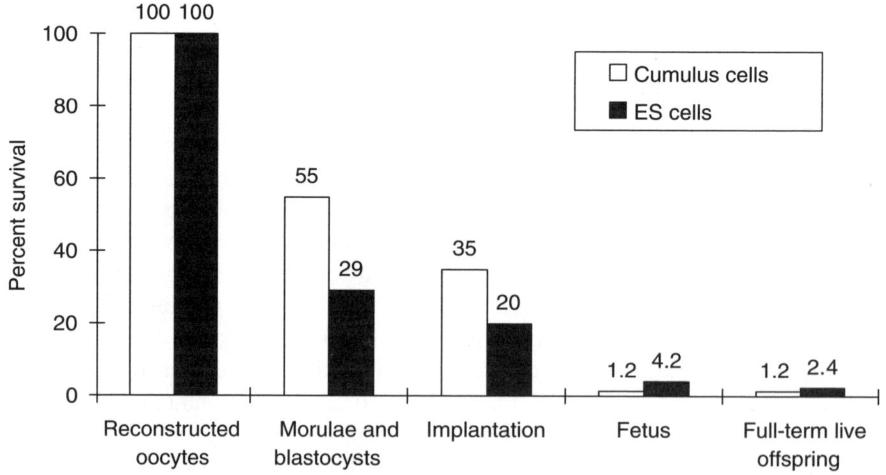

FIGURE 1.1 Survival of cloned embryos at various stages of development. The number of oocytes that survived nuclear transfer of cumulus cells or embryonic stem (ES) cell nuclei is represented as 100%. Advancement to the morula/blastocyst stage occurred over 3.5 days *in vitro*. At 19.5 d.p.c., cesarean section is performed and the number of implantation sites, stillborn fetuses, and live fetuses are noted. The "fetus" stage includes stillborn and live fetuses. (Adapted from Wakayama, T. et al., *Proc. Natl. Acad. Sci. USA*, 1999. 96(26): 14984–14989.)

FIGURE 1.2 Placental hypertrophy in clones. Typical term placentas in cloned mice (left) are two to three times heavier than those of control mice (right).

1.5 PLACENTAL ABNORMALITIES

Placentomegaly is one striking and consistent characteristic of cloned mice [26,31–34] (Figure 1.2). In cloned mice, placentas are enlarged by approximately two- to threefold over that of controls, irrespective of gender of clone, nuclear donor source, or nuclear transfer protocol [26,28]. Abnormal placentas have also been noted in cattle [35] and are thus not unique in cloned mice.

Placental abnormalities in humans have been associated with adverse effects on growth and development of the fetus. Thus, identifying the pathways and genes involved in normal placentation is of great interest. The mechanisms for larger placental size have been more thoroughly studied in cloned mice and have been attributed to hypertrophy of the basal layer, spongiotrophoblasts, giant trophoblasts, and glycogen cells [32]. In addition, placental zonation is disrupted and is characterized by interdigitation of the labyrinthine–basal layer boundary and disorganization of the labyrinthine layer [32].

Similar placental abnormalities have been observed in mice produced by other *in vitro* micromanipulation techniques (intracytoplasmic sperm injection, ICSI; round spermatid injection, ROSI; aggregation chimera; and pronuclear exchange) [34]. However, only cloned mice exhibited basal layer expansion associated with marked proliferation of glycogen cells, suggesting that nuclear transfer is responsible for these specific characteristics [34].

The mechanisms for placental hypertrophy in cloned mice remain unclear. Recent studies have determined that methylation of *Spalt-like gene 3* (*Sall3*), a gene that plays an important role in nervous system development, was increased and highly correlated with larger placentas in cloned mice; they found that hypermethylation was not dependent upon cloned mouse gender, strain, or donor cell used [36]. The degree of methylation at *Sall3* in mice generated by IVF and ICSI was not different from naturally mated controls. These data suggest that this aberration is specific to clones, is not species specific, and thus may result from the nuclear transfer process.

The expression of some imprinted and nonimprinted genes in placentas and fetuses of cloned mice was compared to control mice produced by IVF. At midgestation (E12.5), placental expression of two imprinted genes (*Meg1/Grb10* and *Peg1/Mest*) and two nonimprinted genes (*Igfbp2* and *Esx1*) was significantly decreased in immature Sertoli cell-derived clones compared to IVF controls [37]. In contrast, cloned fetuses exhibited mRNA expression of the same genes that were within the control range.

When the placentas of cloned mice derived from cumulus cells or immature Sertoli cells were examined at full term, the levels of three imprinted genes (*Peg1/Mest*, *Meg1/Grb10*, and *Meg3/Gtl2*) and four nonimprinted genes (*Igfbp2*, *Igfbp6*, *Vegfr2/Flk1*, and *Esx1*) were lower than those of IVF controls, regardless of donor cell source. Three genes (*Igf2*, *H19*, and *Igf2r*) did not exhibit significant difference in clones; however, it was interesting that the variability between clones appeared high, thus suggesting differences within the clone groups. Therefore, although some clones may have gene expression similar to that of controls, expression between individual clones is still highly variable. Furthermore, although the expression of some imprinted and nonimprinted genes was comparable to that of the control group, the placental sizes were still much larger for clones.

These results must be considered carefully; the control group used in this study was generated by IVF. As mentioned earlier, some of the physical manipulations and exposure to *in vitro* culture media are not without side effects; thus, the IVF "control" group may not be the ideal group to which to compare cloned mice.

1.6 GENOMIC IMPRINTING

Aberrant patterns of DNA methylation and expression of imprinted genes have been documented in cloned embryos and progeny in mice regardless of donor cell type [22,23,36–39]. It is interesting that methylation and imprinted gene expression in extraembryonic tissues also have been found to be susceptible to defects [28,37]. Each cloned animal has a different DNA methylation pattern and the extent of hyper- or hypomethylation varies among individuals [39]. Therefore, it is difficult to conclude whether any single area is more susceptible.

1.7 ABNORMALITIES IN NEWBORN CLONES

Intrauterine death prior to birth, developmental retardation, and umbilical hernia are common in cloned mice [34]. A high percentage of cloned pups succumb to respiratory failure at birth [21,26,40]. The reasons for this are currently unclear. Other developmental defects, such as open eyelids at birth [41] (Figure 1.3), have been noted. Birth weight of newborn clones was not different from that of pups derived by IVF or round spermatid injection (ROSI); however, no STOCK control was

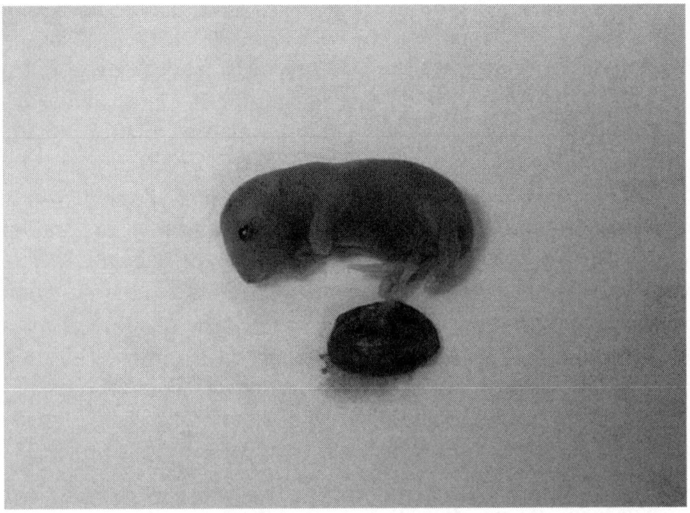

FIGURE 1.3 Developmental defects in clones. Cloned mouse pup born with open eyelids.

included in this study [34]. Increased birth weights of cloned and IVEM mice relative to STOCK control mice [13] have been observed; these observations are consistent with those in cloned and in vitro manipulated cattle and sheep [42].

Despite the high incidence of pre- and perinatal death of cloned mice, some comprehensive studies have been conducted to examine postnatal development, behavior, and phenotype of cloned mice as discussed next. Although ES cells have been used in many nuclear transfer studies, most studies of the long-term consequences of cloning on the phenotype of offspring have been conducted using mice derived from nuclear transfer of somatic cell nuclei. Therefore, this review will focus on cloned mice derived from adult somatic cells.

1.8 PREWEANING DEVELOPMENT OF CLONED MICE

The development of female cloned mice was assessed using the Fox battery of developmental milestones [43]. This battery consists of a set of behaviors, each of which appears at different timepoints throughout neonatal development. No differences were found in these preweaning developmental assays, with the exception of negative geotaxis, ear twitching, and eye opening [44]. Although the appearance of these milestones was delayed in cloned mice compared to controls, the mean day of appearance fell well within the range established for normal mice. Furthermore, the cloned mice performed comparably to control mice in subsequent behavioral tests and were not adversely affected by the delayed appearance of those milestones.

1.9 BEHAVIOR OF CLONED MICE

A well-characterized and widely used task, the Morris water maze, was used to evaluate learning and memory of cloned mice [45]. Clones and IVEM controls successfully completed the task, finding the submerged platform with a shorter latency over consecutive days of testing. Additionally, no differences were observed between the groups, suggesting that both groups were able to use information obtained from previous trials to find the platform faster on subsequent days.

Results also showed that both groups were employing a spatial learning strategy as indicated by the increased amount of time spent in the quadrant where they had been trained to find the platform. Furthermore, when the position of the submerged platform was changed, the cloned mice navigated to the new platform position and found it with a shorter latency than during the initial acquisition trials. Together, these results suggest that cloned mice are capable of completing a spatial learning task and do not have deficits in learning and memory, at least through 3 months of age [44].

Cloned mice have normal diurnal activity patterns as measured by home cage activity that does not differ from that of IVEM control mice at any of the time points examined [44]. Additionally, motor skills and abilities of the clones were assessed and no deficits were found in motor coordination, muscle strength, or balance [44]. In sum, cloned mice appear to develop reflexes and other behaviors at the same ages as normal mice and to have normal motor control and coordination.

Taken together, these behavioral data suggest that cloned mice are not significantly different from control mice from birth through 6 months of age. Longitudinal studies of behavior over the entire life span of cloned mice have yet to be conducted to determine whether cloning has long-term consequences on offspring behavior. Indeed, some studies have reported indications of premature aging in cloned sheep [46] and shorter life spans in cloned mice [47]; whether behavioral deficits occur in concert with these observations remains to be determined. Multiple generations of cloned mice have been examined and no significant deficiencies in any of the developmental or behavioral measures described earlier have been noted [31].

1.10 AGING AND LONGEVITY

The use of differentiated adult somatic cells for cloning calls to question the actual age of a cloned animal. Previous studies have examined telomere lengths as an indirect measure of aging in cloned animals. Progressive shortening of telomeres during DNA replication and cell division has been associated with cellular aging or senescence [48]. Sheep were reported to have shorter telomeres suggestive of premature aging [46] and cattle had age-appropriate [49,50] or longer telomeres [51]. Similarly, cloned pigs have telomere lengths comparable to their age-matched controls [52]. The variation of these findings in cloned animals suggests that cloning has different consequences in different species and even within a species.

As discussed earlier, cloned mice do not show behavioral deficits or signs of premature aging [31,44,53]. To further investigate aging in cloned mice, Wakayama

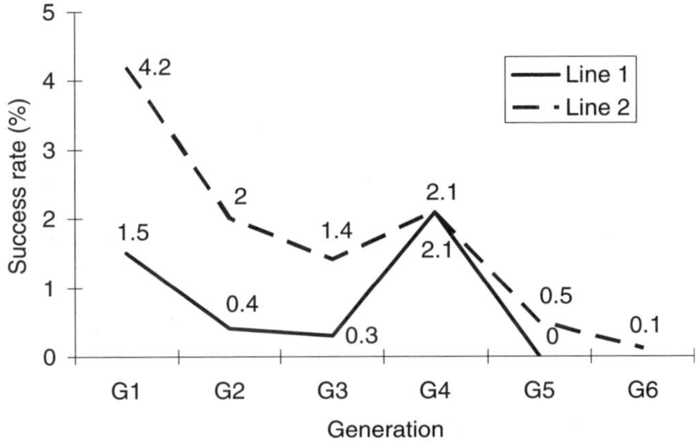

FIGURE 1.4 Success rate of sequential cloning. Successive generations of cloned mice are represented as G1, G2, and so on for two independent mouse lines, A and B. Success rate is expressed as the ratio of the number of live born pups after cumulus cell nuclear transfer to the number of successfully reconstructed oocytes. (Adapted from Wakayama, T. et al., *Nature*, 2000. 407(6802): 318–319.)

et al. successfully generated four and six generations of cloned mice in two independent lines; i.e. clones derived from clones [31] (Figure 1.4).

If mice cloned from adult somatic cells have short telomeres, serial generations of cloned mice should have telomeres that become progressively shorter. Behaviorally, the six generations of cloned mice did not exhibit signs of premature aging. Telomere lengths were also analyzed in this experiment to provide a molecular measure of aging. Consistent with the behavioral results, no telomere shortening was observed. In fact, telomere size increased with each successive generation [31].

A possible explanation for this occurrence involves the enzyme telomerase, which functions to elongate the ends of telomeres and thereby stabilizes and protects the ends of chromosomes during DNA replication and cell division. Although telomerase is not expressed in most somatic cells, telomerase activity was detected in cumulus cells (the donor cells used in these experiments) and suggests that cumulus cells may have longer telomeres to begin with compared to other somatic cells.

In contrast to behavioral and telomere data in cloned mice, Ogonuki et al. reported that male B6D2F1 mice cloned from immature Sertoli cells were prone to early death [47]. The high mortality rate in that study was attributed to higher susceptibility to conditions and diseases, mainly pneumonia and hepatic failure, stemming from reduced immunological function. Immunosuppression has also been reported in cloned cattle [54] and goats [55], suggesting that this condition is not unique to cloned mice. The observations by Ogonuki et al. were restricted to the B6D2F1 mouse strain; 129/Sv and 129 X JF1 Sertoli cell clones did not exhibit the same fate [34], suggesting that early pneumonia-associated death of B6D2F1 Sertoli cell clones was perhaps strain dependent.

Previous studies indicate that telomere shortening may contribute to immunological dysfunctions in mice [56]. Telomere lengths were not measured in the

Ogonuki experiment; however, given the degree of variability found in telomere lengths among cloned cattle [49,51,57,58] (which may be dependent upon the donor cell type), it is plausible that the mice in this particular experiment had shortened telomeres leading to premature immune insufficiency.

The variability of results between these studies suggests that longevity in cloned mice warrants further investigation. The female B6C3F1 cloned mice involved in the behavioral studies described earlier did not die prematurely [53]. In fact, the first cloned mouse, Cumulina, lived for over 2.5 years, a very decent age for a mouse with a 2-year expected lifespan [59]. The longevity of cloned mice may depend on multiple factors, including mouse strain, donor cell source, and age, as well as the technician's skill [19].

1.11 BODY WEIGHT AND OBESITY

Several studies now have reported increased body weight in cloned animals, primarily in mice (Figure 1.5) [13,34,44]. Cloned mice derived from adult cumulus cells have higher body weights and have been found to exhibit characteristics consistent with obesity. Food intake measures showed that cloned mice are not hyperphagic as adults on standard rodent chow [13]. However, whether clones were hyperphagic prior to the onset of obesity at adulthood is not known. Preliminary data also suggest

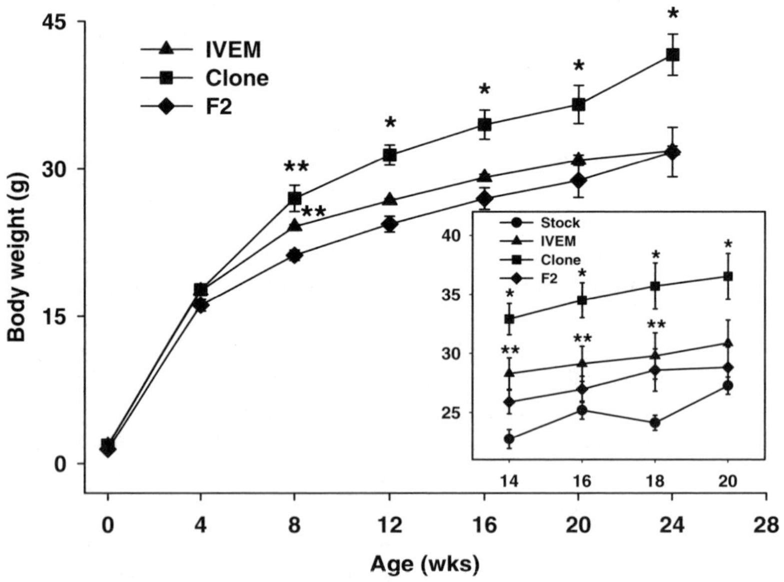

FIGURE 1.5 Body weight of cloned mice, IVEM, and F2 offspring. * $P < 0.05$ vs. IVEM and F2; ** $P < 0.05$ vs. F2. Inset: comparison of body weight of clones, IVEM, and F2 to age-matched background stock mice at 14 to 20 weeks of age. * $P < 0.05$ vs. IVEM, F2, and stock; ** $P < 0.05$ vs. stock. (Adapted from Tamashiro, K.L. et al., *Nat. Med.*, 2002. 8(3): 262–267.)

that clones may have higher metabolic efficiency (weight gain per gram of food consumed) because they do not eat significantly more food per gram of body weight than IVEM or STOCK controls, but continue to gain more weight than controls. Energy expenditure has not been measured in cloned mice and it is possible that it is lower in clones prior to weight gain.

When clones were challenged with an acute hypocaloric challenge, 24-h food deprivation, they lost the same percentage of body weight as STOCK and IVEM controls. When allowed to refeed, clones responded by increasing their food intake similarly to controls, thus indicating that they defend their body weight in a similar fashion [13]. Although adult cloned mice were significantly heavier than their age-matched controls, this does not imply that they are necessarily obese. Body composition analysis revealed that clones had higher body fat compared to controls — almost twice as much; the percentage of lean body mass did not differ among the groups [13]. Consistent with this, cloned mice were also hyperleptinemic and hyperinsulinemic [13]. These data therefore demonstrate that the increased body weight in cloned mice can be associated with an increased percentage of body fat and that cloned mice are indeed obese and exhibit hormonal characteristics consistent with this condition [13] (Figure 1.6).

Examination of possible mechanisms responsible for the development of obesity in cloned mice has begun. Because some animal models of obesity have deficiencies in the melanocortin system, the functioning of this system in adult cloned mice was assessed. Administration of MTII (100 nmol per animal), a synthetic melanocortin agonist, significantly suppressed food intake in cloned mice but not in controls [13]. This result is striking because each animal received the same amount of MTII regardless of body weight. Therefore, because clones are heavier than controls, they received a lower dose of MTII on a milligram per kilogram basis. Cloned mice were also more sensitive to the anorexigenic effect of exogenous leptin (5 μg/g body weight) [13]. These data suggest that the obesity in cloned mice cannot be solely

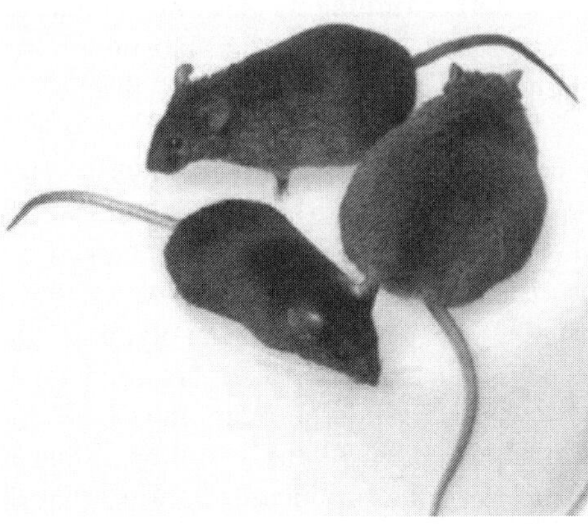

FIGURE 1.6 Cloned and control mice. An adult female B6C3F1 cloned mouse (pictured on the right) and representatives of control groups, STOCK (foreground), and IVEM (top left). (Adapted from Tamashiro, K.L. et al., *Nat. Med.*, 2002. 8(3): 262–267.)

TABLE 1.1
Body Weight and Body Composition of Male B6D2F1 Mice Cloned from Fetal Neurons

	Age-Matched Control (n = 5)	Clone (n = 5)
Terminal body weight (g)	41.4 ± 0.8	56.5 ± 0.9[a]
% Fat tissue	5.9 ± 0.9	16.9 ± 4.1[a]
% Lean tissue	37.6 ± 0.9	39.4 ± 3.5
% Water	56.5 ± 0.9	43.7 ± 0.9[a]

[a] $P < 0.05$.

Note: Data are expressed as mean ± S.E.M.

attributed to deficits in the leptin–melanocortin system. If anything, the clones have a more sensitive melanocortin-signaling system and obesity must be due to other mechanisms.

The majority of the studies summarized thus far have been in female B6C3F1 cumulus cell clones. It is important to point out that increased body weight and obesity in cloned mice is independent of the donor cell type (cumulus cell, fetal neuron, fibroblast, or Sertoli cell); strain of mouse (B6D2F1 and B6C3F1); and gender. Female B6D2F1 cumulus cell clones [13,53] also exhibit the increased body weight phenotype. Male mice are also similarly affected regardless of donor cell source, fetal neuron (Table 1.1) or immature Sertoli cell [34], and strain [B6D2F1 (Table 1.1) or B6129] [34]. Mice that are cloned from ES cells have significantly increased placental and birth weights [40,41]; however, their adult weights were not reported.

Obesity in other species has not been reported; however, the mouse has a much shorter life span compared to larger ruminants and thus increased body weight may not yet be evident in other cloned species. It is also important to point out that laboratory mice are maintained on standard rodent chow throughout the experimental periods, but livestock may be maintained on diets of varying nutrient composition.

Serial cloning of mice may enhance subtle changes resulting from nuclear transfer. To address this possibility, up to six generations of cloned mice were generated [31]. Postnatal development and behavioral characteristics in six generations of clones were not different from controls nor were they different between generations of cloned mice [31]. The obese phenotype is maintained, but not enhanced, in successive generations of mice [31]. Interestingly, when a cloned female was mated with a cloned male or a wildtype male, the obese and placental hypertrophy phenotypes were not passed on to the resulting offspring [13]. Likewise, mating of mice cloned from ES cells produced offspring that did not display the phenotypic abnormalities that their cloned parents displayed, including placental hypertrophy and open eyelids at birth [41]. Together, these data suggest that an epigenetic mechanism may cause aberrant imprinting and/or reprogramming and be responsible for the altered phenotypes [13,41] and that these alterations are corrected during gametogenesis.

1.12 DISCUSSION AND CONCLUSION

Several studies documenting the phenotype of cloned mice have been discussed. The overall conclusion that can be drawn from these studies is that cloning is not without side effects that, in some cases, can be serious enough to threaten and compromise the health and survival of cloned offspring. In addition, some aberrations may be evident immediately in newborn clones; however, the results of several studies now suggest that abnormalities may not be manifest until adulthood. In the mouse, this translates into weeks or months, but in the case of domestic species with longer life expectancies, this spans several years. These studies should ideally be conducted in all cloned species; however, this is not always cost effective and feasible. The mouse is an excellent model to use in exploring the long-term consequences of cloning.

Systematic longitudinal studies of the phenotype of cloned animals must include appropriate control groups. As more investigators include "manipulated" control groups, it is becoming more evident that *in vitro* culture and manipulation alone can produce undesirable side effects in offspring. The effects of these procedures must be studied carefully because the results will have consequences on the use of assisted reproductive techniques as well.

ACKNOWLEDGMENTS

The authors gratefully thank Dr. Hidenori Akutsu, currently at Harvard University, for his assistance with these studies. We also acknowledge Drs. Stephen C. Woods and David A. D'Alessio of the University of Cincinnati for their contributions to the obesity and body weight regulation studies and for valuable discussions about the manuscript. This research was supported by: DK-48061, DK066596 (R.R.S.); the Albert J. Ryan Foundation, NS047791 (K.L.K.T.); the Victoria S. and Bradley L. Geist Foundation; the Kosasa Family Foundation (R.Y.); and the University of Cincinnati Mouse Metabolic Phenotyping Center.

REFERENCES

1. McGrath, J. and Solter, D., Inability of mouse blastomere nuclei transferred to enucleated zygotes to support development *in vitro*. *Science*, 226(4680): 1317, 1984.
2. Wilmut, I. et al., Viable offspring derived from fetal and adult mammalian cells. *Nature*, 385(6619): 810, 1997.
3. Wakayama, T. et al., Full-term development of mice from enucleated oocytes injected with cumulus cell nuclei. *Nature*, 394(6691): 369, 1998.
4. Kato, Y. et al., Eight calves cloned from somatic cells of a single adult. *Science*, 282(5396): 2095, 1998.

5. Polejaeva, I.A. et al., Cloned pigs produced by nuclear transfer from adult somatic cells. *Nature*, 407(6800): 86, 2000.
6. Keefer, C.L. et al., Production of cloned goats after nuclear transfer using adult somatic cells. *Biol. Reprod.*, 66(1): 199, 2002.
7. Shin, T. et al., A cat cloned by nuclear transplantation. *Nature*, 415(6874): 859, 2002.
8. Chesne, P. et al., Cloned rabbits produced by nuclear transfer from adult somatic cells. *Nat. Biotechnol.*, 20(4): 366, 2002.
9. Woods, G.L. et al., A mule cloned from fetal cells by nuclear transfer. *Science*, 301(5636): 1063, 2003.
10. Galli, C. et al., Pregnancy: a cloned horse born to its dam twin. *Nature*, 424(6949): 635, 2003.
11. Zhou, Q. et al., Generation of fertile cloned rats by regulating oocyte activation. *Science*, 302(5648): 1179, 2003.
12. Wilmut, I. et al., Somatic cell nuclear transfer. *Nature*, 419(6907): 583, 2002.
13. Tamashiro, K.L. et al., Cloned mice have an obese phenotype not transmitted to their offspring. *Nat. Med.*, 8(3): 262, 2002.
14. Young, L.E. et al., Epigenetic change in IGF2R is associated with fetal overgrowth after sheep embryo culture. *Nat. Genet.*, 27(2): 153, 2001.
15. Ecker, D.J. et al., Long-term effects of culture of preimplantation mouse embryos on behavior. *Proc. Natl. Acad. Sci. USA*, 101(6): 1595, 2004.
16. Fernandez–Gonzalez, R. et al., Long-term effect of *in vitro* culture of mouse embryos with serum on mRNA expression of imprinting genes, development, and behavior. *Proc. Natl. Acad. Sci. USA*, 101(16): 5880, 2004.
17. Boiani, M. et al., Oct4 distribution and level in mouse clones: consequences for pluripotency. *Genes. Dev.*, 16(10): 1209, 2002.
18. Solter, D., Mammalian cloning: advances and limitations. *Nat. Rev. Genet.*, 1(3): 199, 2000.
19. Perry, A.C. and Wakayama, T., Untimely ends and new beginnings in mouse cloning. *Nat. Genet.*, 30(3): 243, 2002.
20. Young, L.E. and Fairburn, H.R., Improving the safety of embryo technologies: possible role of genomic imprinting. *Theriogenology*, 53(2): 627, 2000.
21. Wakayama, T. et al., Mice cloned from embryonic stem cells. *Proc. Natl. Acad. Sci. USA*, 96(26): 14984, 1999.
22. Humpherys, D. et al., Abnormal gene expression in cloned mice derived from embryonic stem cell and cumulus cell nuclei. *Proc. Natl. Acad. Sci. USA*, 99(20): 12889, 2002.
23. Humpherys, D. et al., Epigenetic instability in ES cells and cloned mice. *Science*, 293(5527): 95, 2001.
24. Gao, S. et al., Effect of cell confluence on production of cloned mice using an inbred embryonic stem cell line. *Biol. Reprod.*, 68(2): 595, 2003.
25. Yamazaki, Y. et al., Assessment of the developmental totipotency of neural cells in the cerebral cortex of mouse embryo by nuclear transfer. *Proc. Natl. Acad. Sci. USA*, 98(24): 14022, 2001.
26. Wakayama, T. and Yanagimachi, R., Cloning of male mice from adult tail-tip cells. *Nat. Genet.*, 22(2): 127, 1999.
27. Ogura, A. et al., Birth of mice after nuclear transfer by electrofusion using tail tip cells. *Mol. Reprod. Dev.*, 57(1): 55, 2000.
28. Ogura, A. et al., Production of male cloned mice from fresh, cultured, and cryopreserved immature Sertoli cells. *Biol. Reprod.*, 62(6): 1579, 2000.

29. Kishikawa, H., Wakayama, T., and Yanagimachi, R., Comparison of oocyte-activating agents for mouse cloning. *Cloning Stem Cells*, 1(3): 153, 1999.
30. Wakayama, T. and Yanagimachi, R., Cloning the laboratory mouse. *Semin. Cell Dev. Biol.*, 10(3): 253, 1999.
31. Wakayama, T. et al., Cloning of mice to six generations. *Nature*, 407(6802): 318, 2000.
32. Tanaka, S. et al., Placentomegaly in cloned mouse concepti caused by expansion of the spongiotrophoblast layer. *Biol. Reprod.*, 65(6): 1813, 2001.
33. Singh, U. et al., Different molecular mechanisms underlie placental overgrowth phenotypes caused by interspecies hybridization, cloning, and Esx1 mutation. *Dev. Dyn.*, 230(1): 149, 2004.
34. Ogura, A. et al., Phenotypic effects of somatic cell cloning in the mouse. *Cloning Stem Cells*, 4(4): 397, 2002.
35. Cibelli, J.B. et al., Cloned transgenic calves produced from nonquiescent fetal fibroblasts. *Science*, 280(5367): 1256, 1998.
36. Ohgane, J. et al., The Sall3 locus is an epigenetic hotspot of aberrant DNA methylation associated with placentomegaly of cloned mice. *Genes Cells*, 9(3): 253, 2004.
37. Inoue, K. et al., Faithful expression of imprinted genes in cloned mice. *Science*, 295(5553): 297, 2002.
38. Mann, M.R. et al., Disruption of imprinted gene methylation and expression in cloned preimplantation stage mouse embryos. *Biol. Reprod.*, 69(3): 902, 2003.
39. Ohgane, J. et al., DNA methylation variation in cloned mice. *Genesis*, 30(2): 45, 2001.
40. Eggan, K. et al., Hybrid vigor, fetal overgrowth, and viability of mice derived by nuclear cloning and tetraploid embryo complementation. *Proc. Natl. Acad. Sci. USA*, 98(11): 6209, 2001.
41. Shimozawa, N. et al., Abnormalities in cloned mice are not transmitted to the progeny. *Genesis*, 34(3): 203, 2002.
42. Young, L.E., Sinclair, K.D., and Wilmut, I., Large offspring syndrome in cattle and sheep. *Rev. Reprod.*, 3(3): 155, 1998.
43. Fox, W.M., Reflex-ontogeny and behavioral development of the mouse. *Anim. Behav.*, 13(2): 234, 1965.
44. Tamashiro, K.L. et al., Postnatal growth and behavioral development of mice cloned from adult cumulus cells. *Biol. Reprod.*, 63(1): 328, 2000.
45. Morris, R.G.M., Spatial localization does not require the presence of local cues. *Learn. Motiv.*, 12: 239, 1981.
46. Shiels, P.G. et al., Analysis of telomere lengths in cloned sheep. *Nature*, 399(6734): 316, 1999.
47. Ogonuki, N. et al., Early death of mice cloned from somatic cells. *Nat. Genet.*, 30(3): 253, 2002.
48. Shay, J.W. and Wright, W.E., Hayflick, his limit, and cellular ageing. *Nat. Rev. Mol. Cell Biol.*, 1(1): 72, 2000.
49. Tian, X.C., Xu, J., and Yang, X., Normal telomere lengths found in cloned cattle. *Nat. Genet.*, 26(3): 272, 2000.
50. Kubota, C., X.C. Tian, and Yang, X., Serial bull cloning by somatic cell nuclear transfer. *Nat. Biotechnol.*, 22(6): 693, 2004.
51. Lanza, R.P. et al., Extension of cell life span and telomere length in animals cloned from senescent somatic cells. *Science*, 288(5466): 665, 2000.
52. Jiang, L. et al., Telomere lengths in cloned transgenic pigs. *Biol. Reprod.*, 70(6): 693, 2004.

53. Tamashiro, K.L. et al., Phenotype of cloned mice: development, behavior and physiology. *Exp. Biol. Med.*, 228(10): 1193, 2003.
54. Renard, J.P. et al., Lymphoid hypoplasia and somatic cloning. *Lancet*, 353(9163): 1489, 1999.
55. Keefer, C.L. et al., Generation of dwarf goat (*Capra hircus*) clones following nuclear transfer with transfected and nontransfected fetal fibroblasts and *in vitro*-matured oocytes. *Biol. Reprod.*, 64(3): 849, 2001.
56. Blasco, M.A., Immunosenescence phenotypes in the telomerase knockout mouse. *Springer Semin. Immunopathol.*, 24(1): 75, 2002.
57. Betts, D. et al., Reprogramming of telomerase activity and rebuilding of telomere length in cloned cattle. *Proc. Natl. Acad. Sci. USA*, 98(3): 1077, 2001.
58. Miyashita, N. et al., Remarkable differences in telomere lengths among cloned cattle derived from different cell types. *Biol. Reprod.*, 66(6): 1649, 2002.
59. Aldhous, P., The cloned mouse that roared is silenced. *Nature*, 405: 268, 2000.

2 Cloning Pigs from Somatic Cells and its Applications

Akira Onishi

CONTENTS

2.1 INTRODUCTION

In the past, it was considered impossible to produce cloned animals derived from differentiated somatic cells. However, since the birth of a cloned sheep [1], successful cloned animals have been reported in mice [2], cattle [3], goats [4], pigs [5,6], cats [7], rabbits [8], mules [9], horses [10], rats [11], and deer (press released). Before cloned animals derived from somatic cells, blastomeres from undifferentiated early stage embryos were mainly used as the nuclear donors. With the animal species mentioned here, embryo clones had been produced from embryos of different early stages.

However, the success rate of embryo clones has been extremely low in pigs. Until recently, there had been only one case of the piglet produced from a blastomere of a four-cell stage embryo [12]. One of reasons for the very low success rate is that intensive research had not been undertaken, mainly because the pig industry did not need to develop embryo-cloning technique for the production of large numbers of genetically superior breeding lines. Another reason is the technical difficulties arising from the unique reproductive characteristics in pig. Unlike in other animals, multiple embryos are needed to establish and maintain pregnancy in sows [13].

Pigs have played an important role in meat production. Recently, new applications of pigs outside animal industry are drawing attention, especially in the biomedical field. Pig organs are expected to be used as human organ replacements in the face of the serious shortage of human organs for transplantation [14,15]. The use of pigs as modeling animals is also gaining ground; for example, they are useful models for studying arteriosclerosis [16]. Thus, the application of pigs in biomedical fields is expanding. To modify pigs desirable to use in biomedical field, a combination of transgenic and cloning technologies is indispensable. This review will discuss the present status of pig cloning and its application.

2.2 CLONING PIGS

2.2.1 Donor Cell Type and Stage of Cell Cycle

Fetal fibroblasts are generally used as donor cells in pig cloning. Cumulus cells and adult cells derived from heart, kidney, oviduct, skin, and ear were also successful for production of cloned pigs. However, it is still unclear which cell type is the most successful for cloning; fetal cells have been preferred because the reverse of chromatin status is believed to be easier in fetal cells than in adult cells.

Quiescent donor cells arrested in G0/G1 phases of cell cycle have been commonly used for production of cloned pigs. Serum starvation and growth arrest when cultured cells reach confluence are usual methods for synchronization in G0/G1 cell cycle stage. When a nucleus at G0 stage is introduced into an enucleated oocyte with high levels of maturation promoting factor (MPF), the nuclear membrane disappears and premature chromosome condensation (PCC) occurs. For maintaining the chromosome in the diploid state, the release of pseudopronuclei out of cytoplasm should be suppressed by substances such as cytochalasin B after activation.

Recently, a cloned pig was obtained by using colchicine-treated somatic cell nuclei as donors [17]; 70.5% in colchicine-treated cells were in G2/M phases. The cell nucleus of the M stage has sister chromatids and the nuclear membrane has already disappeared. Thus, PCC formation by MPF is not necessary. The DNA content of the M stage cell is 4C and half of the pseudopronuclei must be released.

2.2.2 Source of Oocytes

Unfertilized oocytes at the MII stage have been used as recipients of donor nuclei. Usual hormone treatments as injections of equine chorionic gonadotropin (eCG) and human chorionic gonadotropin (hCG) are successful for collection of ovulated oocytes from prepubertal pigs. It is difficult to synchronize estrus in matured pigs by injection of hormones solely, so the oral administration of synthetic progestin, such as Regu-mate or altrenogest, has been combined with subsequent hormone treatment [18]. Regu-mate is effective for synchronization of estrus in matured pigs, but is an unauthorized medicine in Japan. Therefore, another method was chosen for synchronization; this is induced after abortion by injection of prostagladin $F_2\alpha$ analog into early pregnant pigs [19]. This method is very reliable for collection of oocytes and can be adapted to preparation of surrogate mothers for embryo transfer.

For pig nuclear transfer, a large number of oocytes is required; however, direct collection of matured oocytes from pigs is very expensive. Thus, *in vitro* matured oocytes are preferred in many cases [20]. These oocytes are prepared by collecting immature oocytes from abattoir-derived ovaries and allowing them to mature *in vitro*. This method offers the advantage of a stable supply of much larger numbers of oocytes than when they are taken from live animals. In fact, the *in vitro* matured technique of bovine oocytes is well established; therefore, such oocytes are widely used for cloning procedures in cattle. With pigs, however, the developmental potential of oocytes matured *in vitro* is lower than that of *in vivo* matured oocytes [21,22]. Furthermore, because the ovaries are obtained from an abattoir, sufficient precaution must be taken regarding infectious diseases like Aujeszky's disease and porcine respiratory and reproductive syndrome (PRRS).

It is considered that the cytoplasm of the oocytes with high levels of MPF activity is one of the factors inducing reprogramming. When nuclei of somatic cells are transferred into enucleated MII oocytes with high levels of MPF activity, cloned offspring can be obtained. In contrast, activated MII oocytes, in which the activity of MPF was decreased, cannot support the development of embryos reconstructed with somatic cells. MI oocytes before ovulation also have high levels of MPF activity.

It was expected that use of MI oocytes would improve the developmental ability of nuclear transfer embryos in pigs because cytoplasm of MI oocytes was able to expose donor nucleus to high levels of MPF for a longer time than MII oocytes [23]. However, the rate of blastocyst formation in reconstructed embryos with MI oocytes was significantly lower than that of embryos reconstructed with MII oocytes. Other cytoplasmic factors during maturation of oocytes are necessary for maximal development of embryos.

2.2.3 Enucleation

Because it is difficult to observe the nucleus in pig oocytes, removing chromosomes is generally carried out by aspirating the first polar body with a small amount of surrounding cytoplasm. The success rate of this "blind enucleation" method is about 80% in the author's laboratory. Recently, a chemically assisted enucleation procedure was reported in pigs [24]. The treatment of MII oocytes with demecolcin results in a membrane protrusion containing condensed chromosomes. More than 70% of oocytes were able to identify the protrusion and the subsequent enucleation rate was very high (93%) by mechanical removing.

2.2.4 Nuclear Transfer and Activation

Nuclear transfer in pigs can be accomplished by two methods: one is direct injection of donor nuclei into enucleated oocytes and the other is fusion of donor cells with enucleated oocyte by electrical pulses. Both methods can work and deliver live offspring. Timing of oocyte activation is important for the development after nuclear transfer. The matured oocytes, which have already spent some hours by embryo manipulation, are easily activated in pigs. On the other hand, other species, such as cattle oocytes, are difficult to activate by only adding electric pulses and thus chemical activation is generally utilized.

The microinjection method — namely, the introduction of the nucleus into the cytoplasm — can be performed without activation of the oocyte. This enables sufficient exposure of the somatic cell chromosome to the oocyte cytoplasm. In the electrofusion method, if simultaneous fusion and activation of the oocyte occurs, there is the risk of embryogenesis starting before the somatic cell chromosome has sufficient exposure to the oocyte cytoplasm. Polejaeva et al. [6] undertook serial nuclear transfer, in which the first nuclear transfer was conducted by electrofusion of an enucleated unfertilized oocyte and a somatic cell at G0. Subsequently, the nucleus was removed from the first reconstructed embryo and retransferred to an enucleated fertilized oocyte. It can be suggested that the two-stage serial nuclear transfer adequately adjusted the time gap between reprogramming of the nucleus and embryogenesis.

The major disadvantage of the serial nuclear transfer is that the process is very complicated and laborious. At present, usual one-step nuclear transfer has been conducted widely by using electrical pulses for fusion and activation, respectively. Thus, some laboratories prefer using the medium lacking in Ca^{2+} for fusion of cell nuclei into oocytes and subsequent activation was carried out in the medium with Ca^{2+} [24].

2.2.5 Microinjection Method

Figure 2.1 is the flow diagram of the procedure used in the author's laboratory for cloning pigs by the microinjection procedure. The Meishan donor cells that reach confluence are left in the same medium without replacement for about 2 weeks to synchronize in G0/G1 phase. The unfertilized oocytes are collected from the oviducts of hormone-treated Landrace pigs 45 h after injection of hCG. In the case of *in vitro* matured oocytes, the oocytes are matured for 41 to 43 h before enucleation. The enucleation is carried out by aspirating the first polar body with a small amount of cytoplasm into a glass pipette fitted to a piezo-actuated micromanipulator in the medium containing cytochalasin B.

After enucleation, the donor cell is aspirated into a glass pipette with a diameter smaller than the cell to rupture the cell membrane, and the nucleus is directly injected into the cytoplasm of enucleated oocyte. The piezo-actuated micromanipulator greatly improves the efficiency and success rate for enucleation and injection of the nucleus. The reconstructed oocytes are cultured for 1 to 2 h and activated by adding an electric pulse in the medium containing Ca^{2+}. To suppress release of the second polar body (pseudopronuclei), activated oocytes are cultured for 2 h in the medium containing cytochalasin B. They are then switched to the normal culture.

After about 40 h of culture, embryos developed at two- to four-cell stages are transferred into the oviduct of surrogates whose estrus has been synchronized by

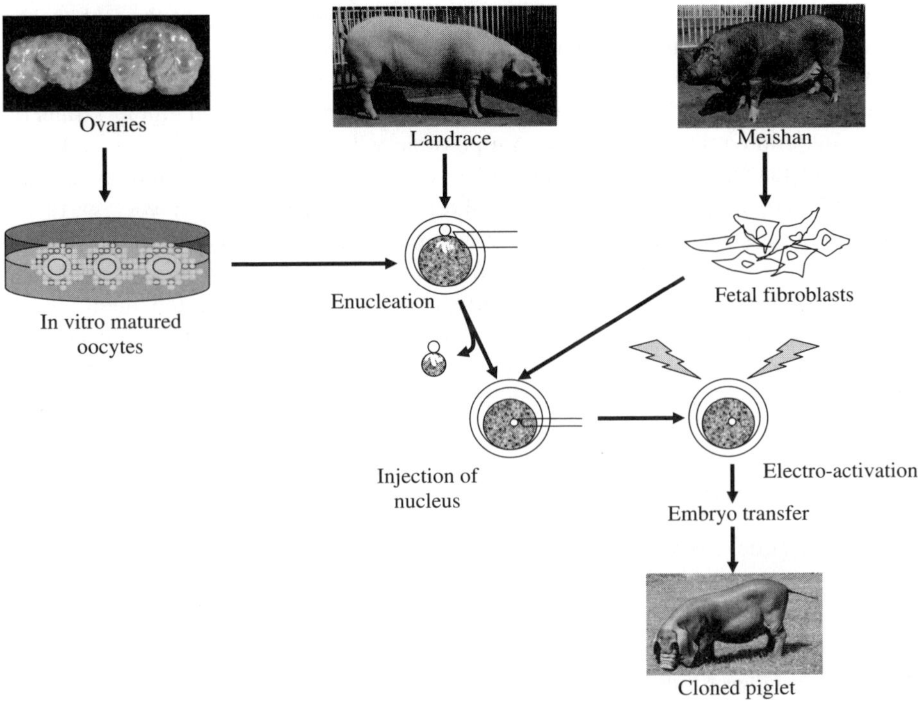

FIGURE 2.1 Pig cloning by microinjection method.

hormone treatment after the induced abortion. Because the nuclear transfer embryos are generally of a low quality, it is recommended that a large number (>100) of embryos be transferred per surrogate. Because at least four fetuses are required for successful pregnancy in pigs, when the number of nuclear transfer embryos is under 30, the author usually cotransfers four fertilized embryos at the same stage for keeping pregnancy.

Many laboratories have tried to repeat the microinjection method in mice, but so far only a few laboratories have succeeded. Mice oocytes are easily damaged by micromanipulation and this weakness is a factor of the difficulty in mice cloning. Conversely, pig oocytes are rather durable for the microinjection method. For example, in mice cloning, the pipette should be filled with mercury or fluorinert for obtaining reliable controls, but an ordinary beveled pipette without mercury is enough for pig cloning. Another advantage of the microinjection is that it requires much shorter time to complete the procedure than the electrofusion method does.

2.3 FUTURE PROSPECTS AND PROBLEMS

2.3.1 Conservation of Genetic Resources

The technology of somatic cell cloning is expected to be applicable to the animal industry, such as increased production of superior breeding stock and conservation of genetic resources. In pigs, it has a great significance as a part of the technology for conserving genetic resources. This is because, unlike in cattle and mice, the cryopreservation of germ cells — embryos in particular — has not been well established in pigs [25]. Therefore, genetic improvements through embryo transfer, as are done in cattle, are currently not possible with pigs.

For instance, the genetic improvement of closed genetic herds must be carried out while preventing the increase of inbreeding coefficient. Pig is very weak for inbreeding depression and the maintenance of closed genetic herds would be ultimately impossible unless a technology for conserving genetic resources is available. Moreover, transgenic pigs have recently become a reality and the methods for conserving them are indispensable. Cryopreservation technology for porcine embryos has definitely shown some advancement in recent years [26]. Somatic cells that can be easily cryopreserved should also be considered for conservation of species.

2.3.2 Transgenic Cloned Pigs

Specific genetic modification by transgenic technology can be applied to various fields. The genetic modification would contribute to improvement of some traits such as growth rate and feed efficiency in pigs. Production of modeling animals to human pathology and biomedical products for human therapy would be possible by transgenes. Thus, a combination of transgenic technology with somatic cell cloning has rapidly started.

The first success of a transgenic cloned animal was reported in sheep [27]. Transgenic cloned pigs were first produced from cultured skin fibroblast cells derived from a 4-month-old H-transferase transgenic male in 2001 [28]. In the same year, a reporter protein: the enhanced green fluorescent protein (EGFP)-expressed cloned pigs were produced from *in-vitro* transfected fetal fibroblast cells [29]. To transduce the fibroblasts, a replication-defective retroviral vector was used. The author's laboratory also established the EGFP-expressed fetal fibroblast cells after transfection with a nonretroviral vector, pCAGGS expression vector. Subsequently, EGFP-expressed pigs were produced by nuclear transfer using the transfected cells (Figure 2.2).

Microinjection of DNA directly into pronuclea of zygote has been used for many years to produce transgenic pigs. However, the efficiency of transferring genes into pigs is quite low — under 1% [30]. Even though transgenic pigs were produced by the microinjection method, variable transgene expression patterns and uncertain transmission into the germ line cannot be avoided. The technology of somatic cell cloning has the advantage in this respect. The integration of gene and level of gene expression can be confirmed and screened in the cultured cells after gene transfer, and it can improve the efficiency of transgenic animals through nuclear transfer using such genetically modified cells. Actually, the author's results showed that all cloned pigs derived from the transfected cells showed expression of EGFP (unpublished data). The certainty of obtaining transgenic animals is especially important for livestock because the period of pregnancy is much longer and the cost per animal is so expensive in livestock compared to mice.

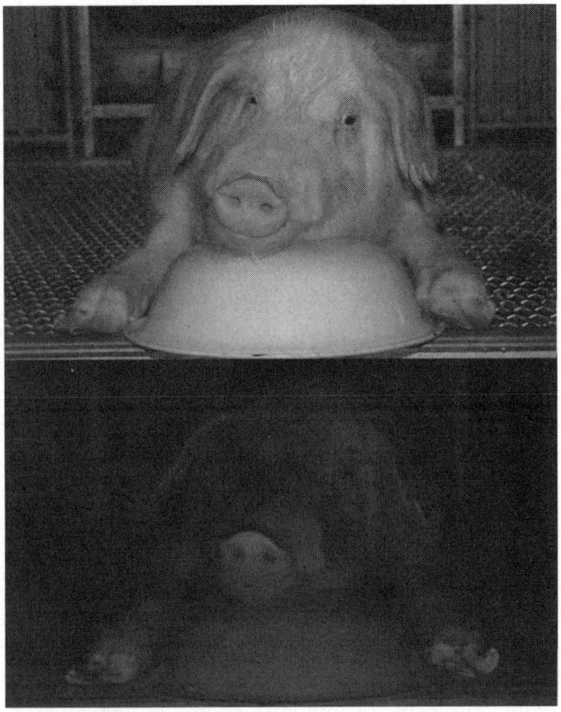

FIGURE 2.2 Transgenic cloned pig showing EGFP expression in the snout and hooves.

The great advantage of production of cloned animals from cultured cells is the opening of the possibility of gene targeting in livestock. Precise gene targeting (mainly for gene disruption by homologous recombination) is achieved in mice using pluripotent embryonic stem (ES) cells. Stable and reliable ES cells are not available in any domestic animals, including pigs, despite considerable efforts to isolate them. The lack of ES cells was a big barrier to gene targeting in livestock.

The first gene-targeted cloned livestock succeeded in sheep inserted with the human α-1-antitrypsis (hAAT) gene into highly expressed but untransrated locus in donor fibroblast cells [31]. hAAT is a candidate protein for human therapy that inhibits inflammatory response caused by excessive destruction of connective tissue. The vector was constructed to be able to produce hAAT in the milk of transgenic sheep. This gene-targeting strategy is to gain a new function by specific insertion of a transgene rather than deletion of endogenous coding sequence. In pigs, it was focused on controlling the antigen–antibody reaction of pig organs for transplantation in humans.

2.3.3 Cloning Pigs for Xenotransplantation

Because organs of pigs are biologically and anatomically similar to human organs, porcine organs are expected to be used as replacements for human organs, which are chronically in short supply. The major problem with xenotransplantation of porcine organs into humans is the hyperacute rejection caused by the reactions involving natural human antibodies and the complement system.

The human decay-accelerating factor (hDAF) gene was transferred into pigs to suppress the hyperacute rejection [32]. When the hearts and kidneys of these transgenic pigs were transplanted into monkeys and baboons, some survived for more than 3 months [33]. However, many rejections occurred and long-term survival cannot be expected. Currently, the work is focused on controlling the antigen–antibody reaction by knocking out the carbohydrate epitope on the surface of porcine cells.

Almost all mammals, with exception of humans, apes, and Old World monkeys, have a galactose-α(1,3)-galactose (α1,3Gal) epitope on the surface of all organisms. It is believed that human natural antibodies against the α1,3Gal epitope mainly cause hyperacute rejection after xenotransplantation. Thus, elimination of the α1,3Gal epitope (encoded by the α1,3GT gene) is expected to extend graft survival of pig organs in primate recipients. Although ES cells are the best for elimination of the α-Gal epitope, these cells have not been established in pigs. In this respect, somatic cell cloning can make major contributions. In 2002, two groups (Immerge BioTherapeutics Inc. and PPL Therapeutic Ltd.) succeeded in producing live cloned piglets in which one allele (heterozygous) of the α1,3GT gene had been disrupted [34,35].

It would take at least 1 year to produce both alleles (homozygous) of α1,3GT knocked-out pigs by natural mating, even if male and female heterozygous pigs might be available at the same time. To speed up the process, both groups attempted to disrupt the second allele by different selection processes. PPL group selected heterozygous fetal cells with toxinA from *Chrostridium difficile,* which has a cytotoxic effect on gal-positive cells [36]. One cell line was selected and survived by a spontaneous T > G transversion at the second allele of the α1,3GT gene. The Immerge group selected heterozygous cells using an antibody against the α1,3Gal epitope [37]. The gal-positive cells are bound with the antibody and destroyed by the following complement treatment. This method relies on the observation that mitotic recombination occurs at a measurable frequency (approximately one in $10^{-4} \sim 10^{-5}$) in primary fibroblasts [38]. Mitotic recombination resulted in the loss of heterozygosity (LOH) at the second allele of the α1,3GT gene.

Subsequently, both groups succeeded in producing α1,3GT-deficient pigs by nuclear transfer. The results of kidney and heart transplantation of α1,3GT-deficient pigs into baboon were reported (oral abstracts) during the seventh congress of the International Xenotransplantation Association held in Glasgow in 2003. The longest survivals in kidney and heart transplantation were 81 days and 62 days, respectively. These results could not exceed the survival period recorded by the conventional hDAF transgenic pigs. Although hyperacute rejection was suppressed by the elimination of the α1,3Gal epitope, delayed xenograft rejection should be solved by further work.

Another factor that cannot be ignored when transplanting porcine organs into humans is infectious disease caused by the porcine endogenous retrovirus [39,40]. The disease does not develop when the virus spreads from one pig to another, but its effect when transmitted to humans is totally unknown. Studies on pig-to-human xenotransplantation would also be important in the light of the serious shortage of human replacement organs for transplantation. However, practical use of porcine organs will be difficult unless the problems related to the endogenous porcine virus are overcome.

2.3.4 Abnormalities in Cloned Pigs

Somatic cell clones of sheep and cattle show high percentages of stillbirths, miscarriages, and malformation of fetuses. They have high postnatal mortality and a higher incidence of placental abnormalities [3,27,42–43]. Pigs cloned from somatic cells have less abnormality. The birth weights, placenta weights, and growth rates were normal in cloned pigs [44]. The author's cloned pigs also showed normal growth rates and reproductive performances (unpublished data). No abnormalities were detected with the immune system of cloned pigs [44].

Sudden death before weaning is very difficult to analyze in cloned pigs because even 5 to 25% of piglets derived from usual mating die between births and weaning on a normal pig farm [45]. However, a high incidence of death was observed in EGFP-expressed cloned pigs. Carter et al. [44] reported that five of ten EGFP cloned piglets died. The author's data showed that seven of eight EGFP-expressed cloned piglets died before weaning (unpublished data). Mice believed to be homozygous for the GFP gene emitted a stronger GFP fluorescence and showed growth retardation and unexplained death before sexual maturation [46].

EGFP, the color variant of GFP, shows greater fluorescence and stability compared with the original GFP and may have a deleterious effect on these animals. To take advantage of these technologies, the technique of cloning pigs needs to become more reliable. Making this technique more precise and reliable is a task for the future.

2.4 CONCLUSION

From the first success of cloned pigs from somatic cells in 2000, it took Immerge BioTherapeutics Inc. and PPL Therapeutic Ltd. only 2 years to produce α1,3GT knocked-out pigs. The recent progress in cloning pigs combined with transgenic technology is remarkable. However, these innovative companies are closed now. It seems that practical use of cloned pigs for the biomedical industry is too early. Further research is needed for the application.

REFERENCES

1. Wilmut, I. et al., Viable offspring derived from fetal and adult mammalian cells, *Nature,* 385, 810, 1997.
2. Wakayama, T. et al., Full-term development of mice from enucleated oocytes injected with cumulus cell nuclei, *Nature,* 394, 369, 1998.
3. Kato, Y. et al., Eight calves cloned from somatic cells of a single adult, *Science,* 282, 2095, 1998.
4. Baguisi, A. et al., Production of goats by somatic cell nuclear transfer, *Nat. Biotech.,* 17, 456, 1999.

5. Onishi, A. et al., Pig cloning by microinjection of fetal fibroblast nuclei, *Science,* 289, 1188, 2000.
6. Polejaeva, I.A. et al., Cloned pigs produced by nuclear transfer from adult somatic cells, *Nature,* 407, 505, 2000.
7. Shin,T. et al., A cat cloned by nuclear transplantation, *Nature,* 415, 859, 2002.
8. Chesne, P. et al., Cloned rabbits produced by nuclear transfer from adult somatic cells, *Nat. Biotechnol.,* 20, 366, 2002.
9. Woods, G.L. et al., A mule cloned from fetal cells by nuclear transfer, *Science,* 301, 1063, 2003.
10. Galli, C. et al., A cloned horse born to its dam twin, *Nature*, 424, 635, 2003.
11. Zhou, Q. et al., Generation of fertile rats by regulating oocyte activation, *Science,* 302, 1179, 2003.
12. Prather, R.S., Sims, M.M., and First, N.L., Nuclear transplantation in early porcine embryos, *Biol. Reprod.,* 41, 123, 1989.
13. Polge, C., Rowson, L.E., and Chang, M.C., The effect of reducing the number of embryos during early stage of gestation on the maintenance of pregnancy in the pig, *J. Reprod. Fertil.,* 12, 395, 1966.
14. Cooper, D.K.C. et al., The pig as potential organs donor for man, in *Xenotransplantation*, Cooper, D.K.C. et al., Eds., Springer, Heidelberg, 1991, 481.
15. Ye, Y. et al., The pig as a potential organ donor for man. A study of potentially transferable disease from donor pig to recipient man, *Transplantation*, 57, 694, 1994.
16. Shimokawa, H. et al., Coronary artery spasm induced in atherosclerotic miniature swine, *Science,* 221, 560, 1983.
17. Lai, L. et al., Transgenic pig expressing the enhanced green fluorescent protein produced by nuclear transfer using colchicine-treated fibroblasts as donor cells, *Mol. Reprod. Dev.,* 62, 300, 2002.
18. Webel, S.K. and Day, M.N., The control of ovulation, in *Control of Pig Reproduction*, Cole, D.J.A. and Foxcroft, G.R., Eds., Butterworth Scientific, London, 1982, 197.
19. Love, R.J. and Grey, R.G., Early abortion of gilts as a way of synchronising oestrus and improving litter size, *Austral. Vet. J.,* 70(12), 452, 1993.
20. Betthauser, J. et al., Production of cloned pigs from *in vitro* systems, *Nat. Biotech.,* 18, 1055, 2000.
21. Funahashi, H. and Day, B.N., Advances in *in vitro* production of pig embryos, *J. Reprod. Fertil. Suppl., 52*, 271, 1997.
22. Wang, W.H. et al., Morphologic evaluation and actin filament distribution in porcine embryos produced *in vitro* and *in vivo*, *Biol. Reprod.,* 60, 1020, 1999.
23. Miyoshi, K. et al., Improvement in cloning efficiencies may be possible by increasing uniformity in recipient oocytes and donor cells, *Biol. Reprod.,* 68, 1079, 2003.
24. Yin, X.J. et al., Production cloned pigs from adult somatic cells by chemically assisted removal of maternal chromosomed, *Biol. Reprod.,* 67, 442, 2002.
25. Dobrinsky, J.R., Cryopreservation of pig embryos, *J. Reprod. Fertil. Suppl.,* 52, 301, 1997.
26. Nagashima, H. et al., Cryopreservation of porcine embryos, *Nature,* 374, 416, 1995.
27. Schnieke, A.E. et al., Human factor IX transgenic sheep produced by transfer of nuclei from transfected fetal fibroblasts, *Science,* 278, 2130, 1997.
28. Bondioli, K. et al., Cloned pigs generated cultured skin fibroblasts derived from an H-transferase transgenic boar, *Mol. Reprod. Dev.,* 60, 189, 2001.
29. Park, K.-W. et al., Production of cloned swine that express the enhanced green fluorescent protein, *Anim. Biotechnol.,* 12(2), 173, 2001.

30. Pursel, V.G. et al., Expression and performance in transgenic pigs, *J. Reprod. Fertil. Suppl.,* 40, 235, 1990.
31. McCreath, K.J. et al., Production of gene-targeted sheep by nuclear transfer from cultured somatic cells, *Nature,* 405, 1066, 2000.
32. Langford, G.A. et al., Production of pigs transgenic for human decay accelerating factor, *Transplant Proc.,* 26, 1400, 1994.
33. Bhatti, F.N.K. et al., Three month survival of hDAF transgenic pig hearts transplanted into primates, *Transplant. Proc.,* 31, 958, 1999.
34. Lai, L. et al., Production of α-1,3-galactosyltransferase knockout pigs by nuclear transfer cloning, *Science,* 295, 1089, 2002.
35. Dai, Y. et al., Targeter disruption of the α-1,3-galactosyltransferase gene in cloned pigs, *Nat. Biothechnol.,* 20, 251, 2002.
36. Phelps, C.J. et al., Production of α-1,3-galactosyltransferase-deficient pigs, *Science,* 299, 411, 2003.
37. Kolber–Simonds, D. et al., Production of α-1,3-galactosyltransferase null pigs by means of nuclear transfer with fibroblasts bearing loss of heterozygosity mutations, *Proc. Natl. Acad. Sci. USA.,* 101, 7335, 2004.
38. Shao, C. et al., Mitotic recombination produces the majority of recessive fibroblast variants in heterozygous mice, *Proc. Natl. Acad. Sci. USA.,* 96, 9230, 1999.
39. Paradis, K. et al., Search for cross-species transmission of porcine endogenous retrovirus in patients treated with living pig tissue, *Science,* 285, 1236, 1999.
40. van der Laan, L.J.W. et al., Infection by porcine endogenous retrovirus after islet xenotransplantation in SCID mice, *Nature,* 407, 90, 2000.
41. Wells, D.N. et al., Production of cloned lambs from established embryonic cell line: a comparison between *in vivo*- and *in vitro*-matured cytoplasts, *Biol. Reprod.,* 57, 385, 1997.
42. Hill, J.R. et al., Clinical and pathologic features of cloned transgenic calves and fetuses (13 case studies), *Theriogenology,* 51, 1451, 1999.
43. Kato, Y., Tani, T., and Tsunoda, Y., Cloning of calves from various somatic cell types of male and female adult, newborn and fetal caws, *J. Reprod. Fertil.,* 120, 231, 2000.
44. Carter, D.B. et al., Phenotyping of transgenic cloned piglets, *Cloning Stem Cells,* 4, 131, 2002.
45. Whittemore, C., *The Science and Practice of Pig Production.* 2nd ed., Blackwell Science, London, 2002, 258.
46. Kawakami, N. et al., Green fluorescent protein-transgenic mice: immune functions and their application to studies of lymphocyte development, *Immunol. Lett.,* 70, 165, 1999.

3 Amphibian Nuclear Transfer and Future Directions of Research

J.B. Gurdon

CONTENTS

3.1 BACKGROUND

Amphibians are the animals in which cloning was first achieved in multicellular organisms. The history of cloning up till the present time has been recently reviewed [1]. Briggs and King [2] were the first to produce normal larvae by transplanting nuclei from early embryos into enucleated eggs. Their work suggested that, as cells differentiate, they lose their ability to promote normal development of eggs. However, soon after the work of Briggs and King, experiments with Xenopus showed that even the nuclei of differentiated intestinal cells could elicit the formation of normal adult male and female frogs [3]. This established the principle that cell specialization does not necessarily require any stable loss of genes or genetic activity.

In all nuclear transplantation experiments in amphibia and in mammals, it has been found that, as cells specialize, their nuclei become less efficient at being successfully transplanted to enucleated eggs to yield normal development. The reason for this is still not known. The two major conclusions from the first half-century of nuclear transplantation are that the genome is conserved during cell differentiation and that egg cytoplasm can reprogram gene expression in somatic cell nuclei.

Soon after nuclear transplantation was successful in amphibia, interest arose in trying to decipher the mechanisms that might be responsible for the remarkable changes that take place. As assays for gene expression were developed, it became clear that transplanted nuclei adopt, very largely and very quickly, the pattern of gene expression characteristic of early embryos. This was true for major classes of gene expression such as ribosomal RNA and tRNA and, of course, for the frequency of DNA synthesis [4]. Only more recently, when quantitative assays of individual gene expressions can be carried out, has it become clear that nuclear transfer embryos often misexpress genes quantitatively [5]. Immediately after transplantation to egg cytoplasm, somatic cell nuclei undergo an enormous enlargement and dispersal of chromatin, coming thereby to resemble egg and sperm nuclei after fertilization.

Because the genome is conserved during development, it must be possible, in principle, to reverse the epigenetic state of a somatic nucleus. The key advance required is to understand the molecules and mechanisms of reprogramming. If this can be done, we can look forward to the prospect of efficiently generating rejuvenated cells for replacement.

3.2 DIRECTIONS FOR FUTURE RESEARCH

3.2.1 Oocytes as Recipient Cells

It is important to appreciate that, in amphibia, gene transcription does not start after fertilization until the mid-blastula transition; this is at the 4000-cell stage, 5 h after nuclear transfer. Therefore, the reprogramming of gene expression may not take place until this stage, and recipient eggs (as opposed to blastulae) may not necessarily have any reprogramming activity. It may even be that the 12 cell cycles between nuclear transfer and the mid-blastula stage may be important in diluting out the epigenetic state of a somatic nucleus.

For this reason, oocytes have come to be used as recipients for nuclear transfer. An oocyte is defined as a growing egg cell at the diplotene stage of first meiotic prophase. Multiple somatic cell nuclei can be injected into an oocyte; these nuclei do not divide and neither does the oocyte. It became clear that somatic cell nuclei must be injected into the germinal vesicle of the oocyte.

Under these conditions, injected nuclei undergo enlargement; they increase their rate of RNA synthesis and new genes are expressed. They therefore undergo reprogramming, as they do when transplanted to eggs, but they do so without nuclear division. It is important to know what reprogramming molecules and conditions exist in these recipient cells. Recently, it has been found that mammalian somatic cell nuclei are reprogrammed when injected into Xenopus oocytes because they are induced to express the stem cell marker gene oct4 [5].

Xenopus oocytes seem to offer a favorable opportunity for further analysis [6–8]. Current methods of molecular biology make it possible to analyze the methylation state of the promoters of a gene. An obviously approachable question is whether the mouse oct4 promoter is demethylated when mouse somatic cell nuclei are reprogrammed in oocytes. If so, it is possible that an extract of oocytes could be prepared that will carry out the demethylation *in vitro* and so lead to the identification of a demethylase. If such can be identified, it may become possible to overexpress a demethylase in mammalian somatic cells and thus at least initiate the reprogramming process.

In the long term, it may become possible to reprogram somatic cell nuclei directly with egg or oocyte extracts. This assumes that proliferating cultures of embryonic stem cells can be derived from reprogrammed somatic cells. Research is currently proceeding in finding ways of directing embryonic stem cells into desired directions of differentiation.

3.2.2 Permeabilized Cells

Hakelien [9] described the reversible permeabilization of cultured cells and the entry of fractionated cell extracts into such cells. Remarkably, some of these cells resealed and continued to grow, and at least a temporary change of gene expression was observed. If this procedure can be made to be reproducible, this would open a wonderful opportunity to create reprogrammed cells and to identify the reprogramming molecules. From this point, it might be possible to proceed, as discussed earlier, by making proliferating embryonic stem cells for replacement.

Compared to mammals — the only nonamphibian multicellular organism in which nuclear transplantation has been successful— amphibia have the great advantage of the amount of reprogramming cytoplasmic material available. For example, a mature Xenopus female can contain 25,000 mature oocytes, and each oocyte is 4000 times larger than a mouse (or any other mammalian) egg. A very large number of injected plasmid DNA molecules, up to 10^9, are converted in a single oocyte into actively transcribing chromatin [7]. Furthermore, the promoter region of any desired gene, incorporated into a plasmid, can be mutated. These characteristics make Xenopus oocytes especially suitable for a molecular analysis of nuclear reprogramming.

The major interest in nuclear reprogramming will come from an understanding of this process in mammals and humans. It seems likely that the reprogramming of mammalian somatic cell nuclei and genes in Xenopus oocytes will reveal mechanisms and molecules that apply to mammals, including humans. It will not be difficult to find the human equivalent of Xenopus genes and thus to test the efficacy of reprogramming molecules identified by the Xenopus route.

REFERENCES

1. Gurdon, J.B. and Byrne, J.A., The first half-century of nuclear transplantation, *Proc. Natl. Acad. Sci. USA*, 100, 8048, 2003.
2. Briggs, R. and King, T.J., Transplantation of living nuclei from blastula cells into enucleated frogs' eggs, *Proc. Natl. Acad. Sci. USA*, 38, 455, 1952.
3. Gurdon, J.B., The developmental capacity of nuclei taken from intestinal epithelium cells of feeding tadpoles, *J. Embryol. Exp. Morphol.*, 10, 622, 1962.
4. Gurdon, J.B., Nuclear transplantation in eggs and oocytes, *J. Cell Sci. Suppl.*, 4, 287, 1986.
5. Byrne, J.A., Simonsson, S., and Gurdon, J.B., From intestine to muscle: nuclear reprogramming through defective cloned embryos, *Proc. Natl. Acad. Sci. USA*, 99, 6059, 2002.

6. Belikov, S. et al., Hormone activation induces nucleosome positioning *in vivo*, *Embo. J.*, 19, 1023, 2000.
7. Belikov, S., Gelius, B., and Wrange, O., Hormone-induced nucleosome positioning in the MMTV promoter is reversible, *Embo. J.*, 20, 2802, 2001.
8. Ridgway, P. and Almouzni, G., Chromatin assembly and organization, *J. Cell Sci.*, 114, 2711, 2001.
9. Hakelien, A.M. et al., Reprogramming fibroblasts to express T-cell functions using cell extracts, *Nat. Biotechnol.*, 20, 460, 2002.

4 Cloning in Cattle

Yukio Tsunoda and Yoko Kato

CONTENTS

4.1 OVERVIEW

The main purpose for developing cloning technology is to increase the efficiency of embryo transfer technology. Artificial insemination and embryo transfer technologies were developed for cattle breeding. Because ejaculated bovine semen can be diluted 200 times per ejaculation and semen is usually collected twice a week, more than 20,000 cows can be artificially inseminated each year by one bull.

Compared with the high efficiency of artificial insemination, the efficiency of embryo transfer is limited because transferable embryos recovered from superovulated cows are limited to approximately 20 per year. Thus, scientists engaged in animal reproduction have developed various new technologies such as twinning, *in vitro* maturation and fertilization, and freezing and sexing of embryos to increase the efficiency of embryo transfer technology. Cloning is one such technology. This chapter reviews past and present achievements and problems, mainly in cattle, but also in other mammals.

4.2 CLONING BY EMBRYO SPLITTING

4.2.1 Historical Aspects

Willadsen [1] reported the first identical twins produced by splitting precompaction sheep embryos. Because the *in vitro* culture of sheep embryos at the two- to eight-cell stage had not been established, Willadsen [1] developed a technique in which a pair of separated blastomeres was inserted into an empty zona pellucida, embedded in agar chips, and then temporarily cultured in ligated sheep oviduct. Several days later, embryos that survived in agar chips were recovered and embryos that developed into morula or blastocyst stages were transferred into final recipients to produce identical twins. Such procedures were also used to produce identical twins in bovine [2].

Micromanipulation of precompaction embryos is laborious and temporary recipients are necessary; therefore, simplified procedures were developed [3]. Monozygotic twin bovine, sheep, goats, and rabbits have been successfully produced by bisecting compacted morulae or blastocysts into halves using a fine glass needle or metal blade [4]. The bisected halves are transferred directly to recipients shortly after *in vitro* culture or after freezing and thawing.

4.2.2 Success Rate and Health Consequences

4.2.2.1 Halved Precompaction Embryos

To split precompaction embryos into two halves, the zonae pellucidae of the embryos are first mechanically removed and then blastomeres are separated into two groups at a high success rate. Because the zona pellucida is necessary for the development of 2- to 16-cell stage embryos to young, blastomeres are inserted into empty zonae pellucidae, embedded in agar chips, and then cultured, mainly in sheep oviducts. The recovery rate of embryos in agar chips is not usually high (40 to 91%, Table 4.1). The potential of separated blastomeres to develop into morulae or blastocysts

TABLE 4.1
Developmental Potential of Embryos Halved after *In Vivo* Culture

		Recovery from Temporary Recipients		No. of Fetuses or Young/No. of Half Embryos Transferred (%) [Sets of Twin]		
Species	Embryonic Stage	Recovery Rate	Developmental Rate to Morula/Blastocyst	Fetuses	Young	References
Mouse	2-cell	40	70–91	–	23/80 (29) [6]	4
Rat	2-cell	77–81	68–78	–	43/154 (28) [9]	5
Sheep	2-cell	65	88	–	15/32 (47) [5]	1
Goat	2–4-cell	75	76	–	6/14 (43) [2]	6
Bovine	7–60-cell	91	92	21/28 (75) [10]	–	2
Horse	2–8-cell	86	–	–	5/10 (50) [2]	7

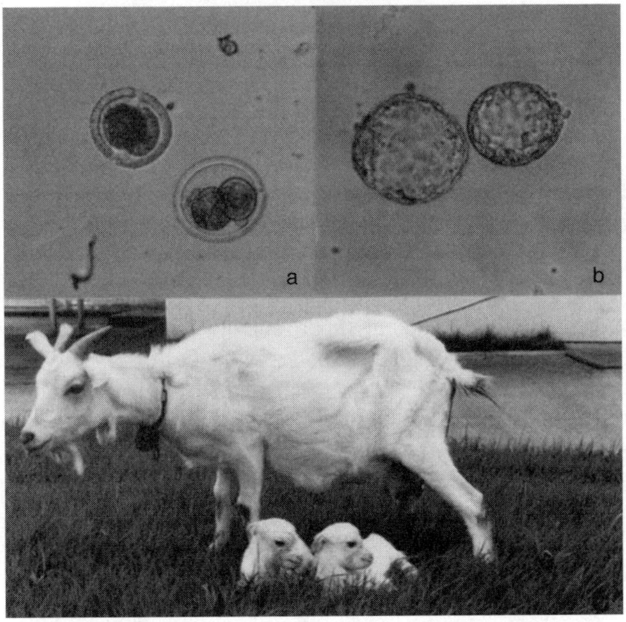

FIGURE 4.1 Production of monozygotic goat twins: a. set of twin embryos embedded in agar; b. set of twin blastocysts released from agar after culture in goat oviduct; c. set of twins. (From Tsunoda, Y. et al., *Jpn. J. Zootech. Sci.*, 55, 643, 1984. With permission.)

is high and 28 to 50% of them develop to term (Figure 4.1) [1–7]. To date, clear abnormalities or perinatal death of live bovine has not been reported.

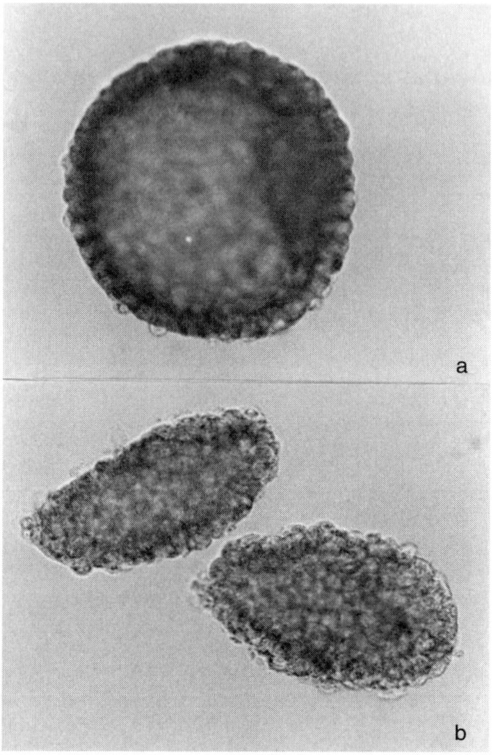

FIGURE 4.2 Bisection of day ten bovine blastocysts: (a) before bisection and (b) after bisection.

4.2.2.2 Halved Postcompaction Embryos

Embryos recovered from the uterus of superovulated cows can be cut into two halves (Figure 4.2). Before bisection, the zonae pellucidae of embryos are removed in some cases and bisected embryos are directly transferred to uteri of recipients. The success rate for bisection is high and identical twins have been produced by embryo bisection in rabbit, sheep, goat, pig, and bovine [8–16]. In some cases, the potential of bisected embryos to develop into live young was high (Table 4.2). Arave et al. [15] reported a large-scale embryo transfer study of bisected bovine embryos. They transferred 181 bisected embryos produced from 91 embryos; 92 of them (51%) produced calves, including 26 identical twins. The pregnancy rate on day 60 after transfer of bisected embryos was 57%, but 6% of pregnant females aborted before parturition.

4.2.2.3 Splitting into Fourths or More

Although monozygotic triplet heifer calves have been successfully produced by separating an eight-cell embryo into four pairs and then culturing in ligated sheep oviduct before transfer to recipients [2], the success rate is low. One healthy rhesus monkey was produced from a quarter of an eight-cell embryo [17] and calves were produced by bisecting a morula into quarters [18]. The production of quadruplets

TABLE 4.2
Developmental Potential of Half Embryos Bisected at the Morula or Blastocyst Stage

		No. of Fetuses or Young/No. of Half Embryos Transferred (%) [Sets of Twin]		
Species	**Embryonic Stage**	**Fetuses**	**Young**	**References**
Rabbit	Morula	–	31/162 (19)	8
Sheep	Morula–blastocyst	–	8/18 (44) [8]	9
	Morula–blastocyst	–	16/34 (47) [7]	10
Goat	Blastocyst	–	9/20 (45) [4]	11
Pig	Morula–blastocyst	–	14/79 (18)	12
	Morula–blastocyst	–	126/759 (17) [7]	13
Bovine	Morula	15/28 (54) [6]		3
	Morula–blastocyst	–	7/28 (25) [3]	14
	Morula–blastocyst	–	92/181 (51) [26]	15
	Morula		3/20 (15)	16
	Blastocyst		10/22 (46)	

might be more difficult compared with splitting precompaction embryos because cells at the bisecting plane are likely to be damaged. If an eight-cell sheep embryo is separated into single blastomeres and cultured in a ligated sheep oviduct, the potential of one eighth of the embryo to develop into young is very low [19].

Because the timing of blastocyst formation is fixed for each species, a blastomere of an eight-cell embryo will form blastocysts with no or a small inner cell mass at the eight-cell stage after cell division three times. To overcome a small inner cell mass cell number, Willadsen and Fehilly [20] aggregated a single blastomere of an eight-cell sheep embryo with a four-cell embryo in which one blastomere had been previously removed. This manipulation was based on the finding that the fast developing blastomere had a better chance to be located in the inner cell mass [21]. Willadsen and Fehilly [20] succeeded in producing five identical sheep. Triplet mice were also produced by aggregating a blastomere of an eight-cell embryo with a parthenogenetic four-cell embryo [22]. Cloning using similar procedures has not been performed in bovine.

4.2.2.4 Application under Commercial Situation

Splitting of bovine embryos has been used under commercial field conditions to increase the production of calves by embryo transfer [23–25] (Table 4.3). Although the final outcome is not clear, the pregnancy rates of bisected demi-embryos on days 50 to 90 were 50 to 59%, which is not different from that after transfer of intact embryos. Leibo and Rall [23] collected 937 day-6 or day-7 embryos from 186 superovulated cows; 422 embryos were bisected into halves but 515 embryos were transferred without bisection. They obtained 441 pregnancies after transfer of 842 demi-embryos (52%) and 291 pregnancies after transfer of 515 intact embryos

TABLE 4.3
Splitting of Bovine Embryos under Field Conditions

Group	No. of Recipients Pregnant/Transferred (%)	References
Demi-Embryo	441/842 (52)	23
Intact	291/515 (57)	
Demi-Embryo	78/132 (59)	24
Demi-Embryo	997/1988 (50)	25

(57%). Thus, they obtained a total of 732 pregnancies; this was 203 more pregnancies than if none of the 937 embryos had been bisected.

4.2.2.5 Combination with Freezing Technology

Freezing demi-embryos increases embryo efficiency. Freezing and splitting procedures — especially bisection of postcompaction embryos — damage embryos somewhat; therefore, the potential of demi-embryos to develop into young after freezing and thawing is low. The main reason for the low viability of frozen–thawed embryos is due to the lower cell number of embryos as well as to injury caused by the fracture planes formed in the ice crystal matrices that directly touched the section of demi-embryos. The insertion of demi-embryos into empty zonae pellucidae improves embryo viability [26].

4.2.2.6 Health Consequences

Only a few studies have examined gestation period, body weight, abnormalities, and peri- and postnatal death of young derived from demi-embryos (Table 4.4). The body weight of day 18 mouse fetuses developing from demi-embryos was significantly lower than that of control fetuses [27]. The body weight of kids from demi-embryos was within the normal range of kids produced by natural mating [11].

TABLE 4.4
Body Weight of Young Derived from Demi-Embryos

Species	Group	No. of Young Examined	Average Body Weight ±SD	References
Mouse	Control	60	1.17 ± 0.20 g	27
	Demi	31	1.04 ± 0.23 g	
Goat	Control		1.59 ± 0.39 kg	11
	Demi	12	1.76 ± 0.21 kg	

Although demi-embryos can increase their cell number before implantation, the time interval might be insufficient in the mouse because the implantation window is narrow compared with goats.

Reichelt and Niemann [13] reported that the loss of postparturition piglets obtained from bisected embryos was higher than that of piglets produced after natural mating. Abnormalities such as cryptorchism, atresia, and hernia formation were high in piglets that developed from bisected embryos. There are no reports of body weight and health consequences in calves derived from demi-embryos.

4.3 CLONING BY NUCLEAR TRANSFER OF EMBRYONIC CELL NUCLEI

4.3.1 Historical Aspect

Because the maximum number of identical calves obtained by embryo splitting is limited to two in most cases, the development of nuclear transfer technology was desired. A reliable technique for nuclear transfer in mammals was first reported in 1983 by McGrath and Solter [28], who exchanged pronuclei between different mouse zygotes and obtained live pups after transfer of nuclear-exchanged zygotes. Soon after this success, the potential of nuclei to develop from the two-cell to blastocyst-stage embryos was tested by introducing the nuclei into enucleated zygotes or two-cell embryos.

The developmental potential of nuclear-transferred eggs, however, was limited [29,30]. Willadsen [31] reported the production of a normal lamb derived from nuclei of eight-cell embryos by using enucleated unfertilized oocytes at the second metaphase (MII), instead of zygotes, as recipient oocytes. Soon after that, normal calves were also produced by nuclear transfer of 2- to 32-stage embryos [32].

4.3.2 Success Rate

Nuclei transferred to enucleated zygotes or two-cell embryos do not lose their developmental memory [30], but nuclei transferred into enucleated oocytes function as if they had been changed into nuclei just after fertilization [32]. Such a change in nuclear memory is called nuclear reprogramming. In earlier studies, oocytes recovered from superovulated cows were used as recipient oocytes and nuclear-transferred oocytes were cultured in ligated oviducts.

Current studies on bovine cloning are performed using *in vitro* systems. A donor cell (blastomere) from precompaction embryos is fused by electric pulses with enucleated oocytes at the MII stage matured *in vitro* whose chromosomes were previously removed (Figure 4.3 and Figure 4.4). Fused oocytes are parthenogenetically activated and cultured *in vitro* for 7 to 9 days before transfer to recipients or frozen storage. The success rate of nuclear-transferred oocytes to develop into blastocysts *in vitro* and into young after transfer to recipients is largely different, depending on the experimental conditions; 10 to 40% of nuclear-transferred oocytes develop to blastocysts and 10 to 30% of transferred blastocysts develop into calves.

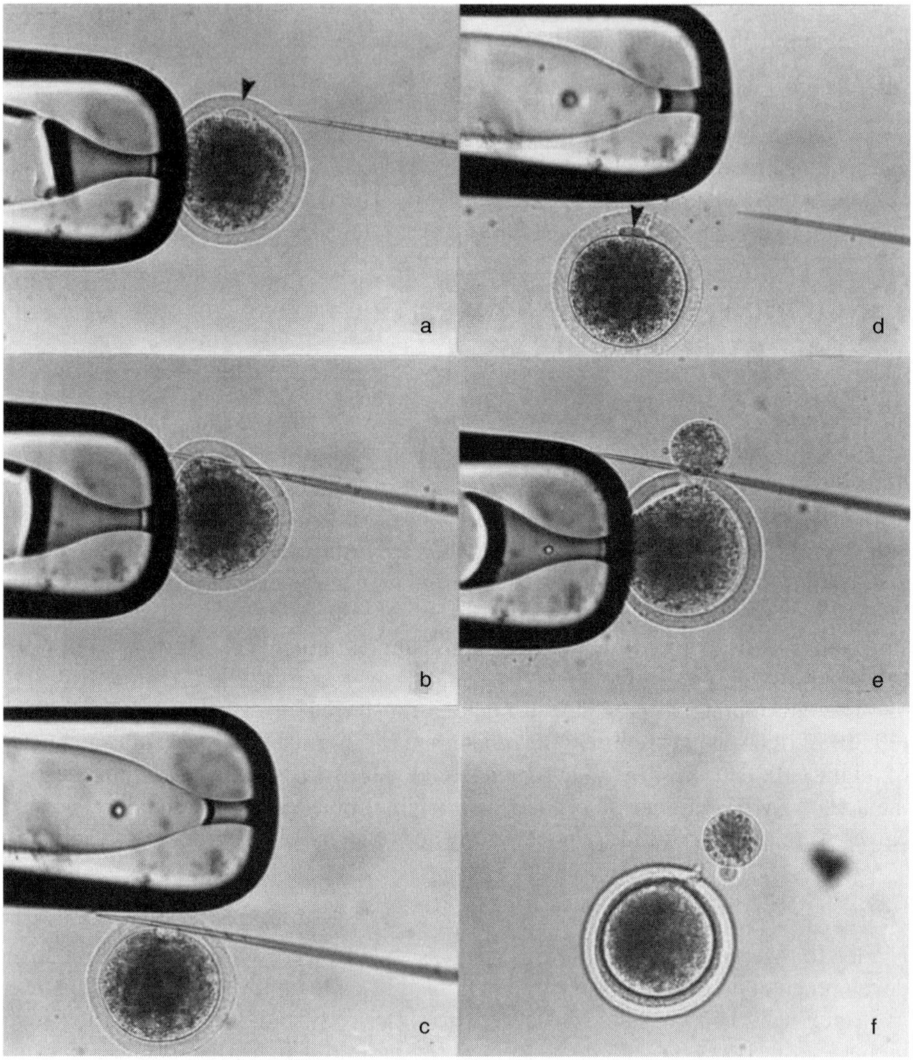

FIGURE 4.3 The procedures for removing the metaphase chromosomes in bovine oocytes: a. and b. the oocyte is sucked with a small bore holding pipette (left) and a fine glass needle (right) is inserted into the pervitelline space near the first polar body (arrow); c. and d. the needle is rubbed against the wall of the pipette to tear the zona pellucida; e. the oocyte is pushed from above using a glass needle and a small amount of cytoplasm near where the first polar body was pushed out from the slit on the zona pellucida; f. a small volume of cytoplasm with the first polar body removed from an oocyte is stained with Hoechst. When metaphase chromosomes are confirmed to be present in the cytoplasm, the rest of the oocyte cytoplasm is used as the recipient cytoplasm. (Modified from Tsunoda, Y. and Kato, Y., *Zool. Sci.*, 17, 1177, 2000.)

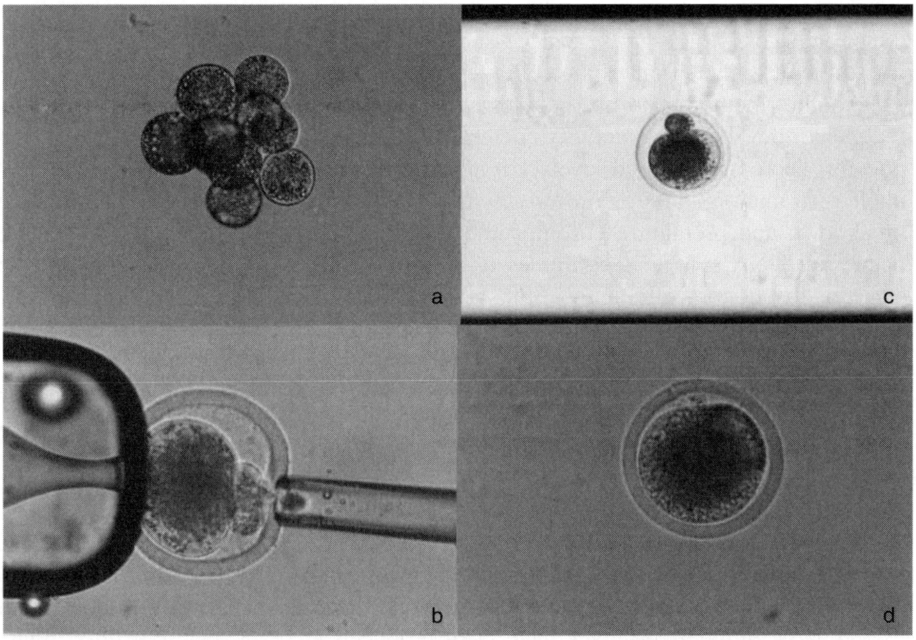

FIGURE 4.4 Nuclear transfer of a blastomere from an 8- to 16-cell bovine embryo: a. zona pellucida-free 8- to 16-cell bovine embryo; b. a blastomere is inserted into the pervitelline space of an enucleated oocyte; c. oocyte–blastomere pair is placed in a 1-mm stainless steel electrode chamber overlaid with fusion medium. Two sets of current pulses are applied several times. d. Fused oocyte.

4.3.2.1 Cell Cycle Stage

One important factor that affects cloning efficiency is the combination of cell cycle stage between donor cells and recipient oocytes [33]. The important key is the presence of maturation promoting factor (MPF), whose activity is high in MII oocytes but declines soon after parthenogenetic activation. Because young oocytes, in which MPF activity is high, are not easily activated parthenogenetically, earlier studies on bovine cloning used aged oocytes in which MPF activity is low as recipient cytoplasm. Recent studies, however, revealed that parthenogenetic activation is successfully induced in young oocytes if they are treated with electric stimulation or ethanol treatment in combination with cycloheximide or 6-dimethyl aminopurine [34].

Nuclei of donor cells introduced into enucleated MII oocytes show nuclear membrane breakdown and premature chromosome condensation. After parthenogenetic activation, the nuclear membrane reforms and DNA synthesis begins, irrespective of the cell cycle stage of the donor nucleus. Thus, the cell cycle stage of the donor cells before nuclear transfer should be G1 or G0. If nuclei at the G2 or M phase are introduced, nuclear-transferred oocytes should release the second polar body after activation. If nuclei at the S phase are used, the chromosome constitution of the nuclear-transferred oocytes becomes abnormal. Because the G1 stage in the cell cycle of preimplantation embryos is short, synchronization of the cell cycle of donor embryos into G1 phase is necessary. Unlike in mouse embryos [35], however, it is difficult to synchronize the cell cycle of bovine preimplantation embryos.

Another method is to use oocytes activated 6 to 9 h before nuclear transfer as recipient oocytes. MPF activity, determined by cdc2 kinase activity in bovine oocytes, decreases to basal levels 1 h after parthenogenetic activation and the pronuclear formation starts 5 h after activation [36]. In this case, nuclear membrane breakdown and premature chromosome condensation of the donor nuclei do not occur, but DNA synthesis continues or stops depending on the original cell cycle stage of the donor cells at the time of nuclear transfer. Normal development of nuclear-transferred oocytes is expected, irrespective of the cell cycle stage of the donor nucleus. Thus, this method is more practical for cloning bovine and a large number of calves have been produced using it. The chromosomes of donor cells are not directly exposed to recipient oocyte cytoplasm. However, they might be reprogrammed during nuclear swelling of the donor cells, although the extent of swelling is less than in MII oocytes [37,38].

4.3.2.2 Donor Cell Type

Since the first successful production of nuclear-transferred calves by Prather et al. [32], a large number of calves have been produced after nuclear transfer of blastomeres of preimplantation stage embryos, inner cell mass cells with or without *in vitro* culture, and cultured cells of genital ridges (Table 4.5) [32,37–42]. Willadsen et al. [39] demonstrated the effectiveness of nuclear transfer technology for cloning a super elite cow. They obtained morulae and blastocysts by fusion of enucleated oocytes with blastomeres from 8- to 64-cell stage embryos following culture in sheep oviducts. Of the 302 recipients that received one nuclear-transferred embryo, 100 recipients (33%) produced calves in a year.

TABLE 4.5
Key Reports in Bovine Cloning by Nuclear Transfer of Embryonic Cells

Donor Cell Stage	No. of Recipients (%) No.	Pregnant	Calved	No. of Calves/No. of Embryos Transferred (%)	References
2–32	13	5 (38)	2 (15)	2/19 (11)	32
8–64	302	128 (42)	100 (33)	100/302 (33)	39
Inner cell mass	15	4 (27)	2 (13)	2/15 (13)	37
	26	6 (23)	4 (15)	4/26 (15)	38
Cultured inner cell mass	27	13 (48)	4 (15)	4/34 (12)	40
Cultured genital ridge cell	73	39 (53)	14 (19)	14/73 (19)	41
	94	50 (53)	20 (21)	20/94 (21)	42

TABLE 4.6
Effects of Age of Donor Embryo on the Developmental Potential of Bovine Nuclear-Transferred Oocytes

Age of Donor Embryo (Day After *In Vitro* Fertilization)	No. of Cells in Donor Embryo (Mean ±SD)	No. of Oocytes Fused/Examined (%)	No. of Oocytes Developed to (%)		
			2-Cell	8-Cell	Blastocyst
3.5	13.2 ± 2.2	167/203 (82)	136 (81)[a]	56 (34)	16 (100)
4.5	17.0 ± 2.8	109/126 (87)	86 (79)	37 (34)	12 (11)
5.5	27.1 ± 1.2	161/184 (88)	110 (68)[b]	49 (30)	15 (9)
6.5	35.0 ± 9.2	132/169 (78)	98 (74)	40 (30)	8 (6)

Note: Values with different superscripts differ significantly ($p<0.05$).

Source: Modified from Ohkoshi, K., Hata, M., Kato, Y., and Tsunoda, Y., *J. Reprod. Develop.*, 43, 261, 1997.

Because the procedures for nuclear transfer are slightly different among reports, precise comparison of the potential of nuclear-transferred oocytes to develop into blastocysts is difficult. The potential of nuclear-transferred oocytes to develop into blastocysts is not significantly different based on the age of donor embryos obtained from *in vitro* fertilization (Table 4.6) [43]. Westhusin et al. [44] reported that the age of donor embryos between day 5 and day 6 did not affect the potential of nuclear-transferred oocytes to develop into morulae and blastocysts in ligated sheep oviducts.

Nuclear-transferred oocytes receiving inner cell mass cells developed into blastocysts and also to live calves after transfer of nuclear-transferred blastocysts into recipients [37,38]. Collas and Barnes [37] reported two calves, but Keefer et al. [38] obtained four calves, two of which were stillborn and two that died shortly after birth. Although embryonic stem cells have not been established in bovine, live calves were produced after transfer of nuclear-transferred oocytes receiving short-term cultured inner cell mass cells (Table 4.5). Sims et al. [40] obtained four calves but two of them were stillborn.

The enucleated oocytes receiving day 12.5 to day 16.5 male mouse fetal germ cells developed into blastocysts but not into offspring [45]. The failures are probably related to the starting point of gamete imprinting for the next generation in male germ cells [46]. One live calf was born after nuclear transfer of the male genital ridge on days 50 to 57 [47], suggesting that the starting point of gamete imprinting might be different. The calf died, however, due to respiration problems 24 h after birth. A large number of cloned calves have been produced after nuclear transfer of transgenic or nontransgenic cultured genital ridge cells obtained from bovine fetuses of 40 to 60 days of age [41,42, Table 4.5]. Forsberg et al. [42] reported that calves were produced after nuclear transfer of presumptive fetal germ cells, but it is not known whether the cultured cells used in these studies were of embryonic or somatic cell origin.

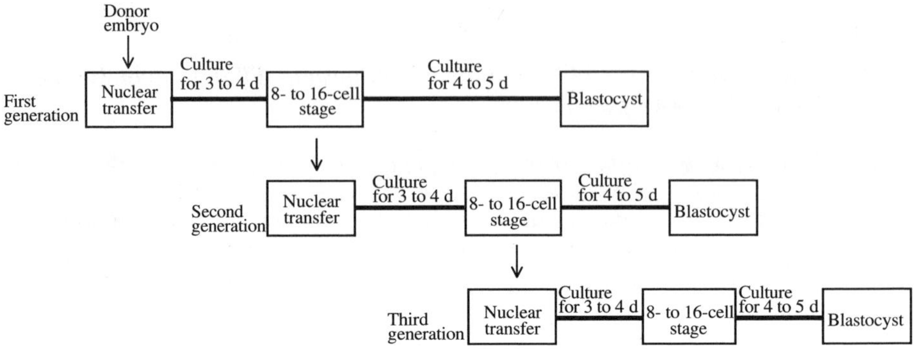

FIGURE 4.5 Procedures for serial nuclear transfer.

4.3.2.3 Serial Nuclear Transfer and Freezing of Nuclear-Transferred Embryos

One method for producing a large number of identical calves is to use cultured embryonic cells. Another method is to use nuclear-transferred embryos as donor cells for the next generation of nuclear transfer [44,48–51]. In one study [51], a single blastomere of one donor embryo at the morula stage was fused with enucleated oocytes by electric pulses. Fused oocytes were cultured *in vitro* for 7 to 9 days, but some reconstituted oocytes that developed to the 8- to 16-cell stage were removed halfway through the *in vitro* culture and used as donor cells for the next generation (Figure 4.5).

Recycling nuclear transfer was again performed. A maximum of 43 or 54 cloned blastocysts were produced from one cow embryo after culture *in vitro* [51] or in sheep ligated oviduct [49]. Cloned blastocysts were obtained even after nuclear transfer three [51] (Table 4.7) or six times [49], and calves were successfully produced after transfer of cloned blastocysts in the second and third generations. As many as ten clonal calves have been produced by serial nuclear transfer [49].

Frozen storage of cloned embryos is required for practical application of nuclear transfer in the animal industry. In the authors' study [51], the *in vitro* viability of

TABLE 4.7
Developmental Potential of Serial Nuclear-Transferred Bovine Embryos

Generation of Nuclear Transfer	No. of Oocytes Developed to Blastocysts/Cultured (%)	No. of Calves/Transferred (%)	
		Fresh	Frozen-Thawed
1	28/133 (21)	2/2 (100)	2/5 (40)
2	65/177 (37)	0/10 (0)	1/4 (25)
3	53/271 (20)	2/8 (25)	0/5 (0)

cloned blastocysts after slow freezing and thawing by using 1.5 *M* ethyleneglycol was low (59%), but not significantly different from that in blastocysts fertilized *in vitro* (72%). As shown in Table 4.7, live calves were also produced from frozen–thawed nuclear-transferred embryos.

4.3.3 Health Consequences

Only a few reports have examined the health consequences in calves produced by nuclear transfer of oocytes receiving embryonic cells. Willadsen et al. [39] reported that 10 of 100 calves produced by nuclear transfer died or were killed due to various reasons. Two died due to unknown reasons a few days after birth. Four calves died during or soon after birth due to hard pulling during assisted birth. The other four had congenital malformations: a large septal defect between the left and right cardiac ventricles; contractures affecting limbs and spine; and facial twist. They also observed that a number of cloned calves were exceptionally large (in excess of 58.9 kg).

Keefer et al. [38] reported that two live calves obtained from nuclear transfer of inner cell mass cells had large birth weight (59.8 and 72.5 kg), and one of them had flexural limb deformity and died from septicemia 15 days after birth. The other had no anatomic abnormalities, but died 4 days after birth for unknown reasons.

Since the first nuclear-transferred calf was born in August 1990 in Japan, 686 calves have been produced in universities and national and prefectural research institutes (from the report of the Japanese Ministry of Agriculture, Forestry and Fisheries, Table 4.8). Of the 686, 123 (17.9%) calves were dead or died before or shortly after birth, but the causes were not reported. The remaining calves went on to be bred, sold, or slaughtered, or died from accidental causes. A number of cattle, however, have calved normally.

TABLE 4.8
The Proportions of Nuclear-Transferred Calves That Died Before or After Birth

Situation	Orgin of Donor Cells	
	Embryonic	Somatic
Total birth	686	367
Stillborn (%)	73 (10.6)	58 (15.8)
Perinatal death (%)	30 (4.4)	50 (13.6)
Death from sickness (%)	20 (2.9)	66 (18.0)

Source: From the report of the Japanese Minstry of Agriculture, Forestry and Fisheries on September 30, 2003.

4.4 CLONING BY NUCLEAR TRANSFER OF SOMATIC CELL NUCLEI

4.4.1 Historical Aspects

In 1994, Collas and Barnes [37] reported that enucleated oocytes receiving cultured granulosa cells developed into blastocysts, but no pregnancy was detected after transfer to recipients. In 1997, Wilmut et al. [52] reported the first successful production of a cloned sheep, Dolly, by nuclear transfer of enucleated oocytes receiving cultured mammary gland cells. Dolly is the first animal obtained from adult somatic cells. The authors reported the first successful production of cloned calves from adult somatic cells in 1998 [53]. Since then, a large number of cloned calves from fetal and adult somatic cells have been produced.

4.4.2 Success Rate

The procedures used for bovine somatic cell nuclear transfer are slightly different among laboratories. In the authors' laboratory [54,55], MII oocytes matured *in vitro* for 20 to 24 h and their chromosomes were mechanically removed and used as recipient cytoplasm. The maternal chromosomes are usually removed by aspiration or by pushing out the cytoplasm overlying the first polar body with or without Hoechst staining (Figure 4.3).

Recently, a new technique for removing chromosomes has been developed [56,57]. Brief treatment of *in vitro* matured oocytes with demecolcine or nocodazole results in a membrane protrusion containing a condensed chromosome mass that can be easily removed by aspiration (Figure 4.6). Donor cells obtained from various tissues are cultured and passaged several times *in vitro* and then induced to the

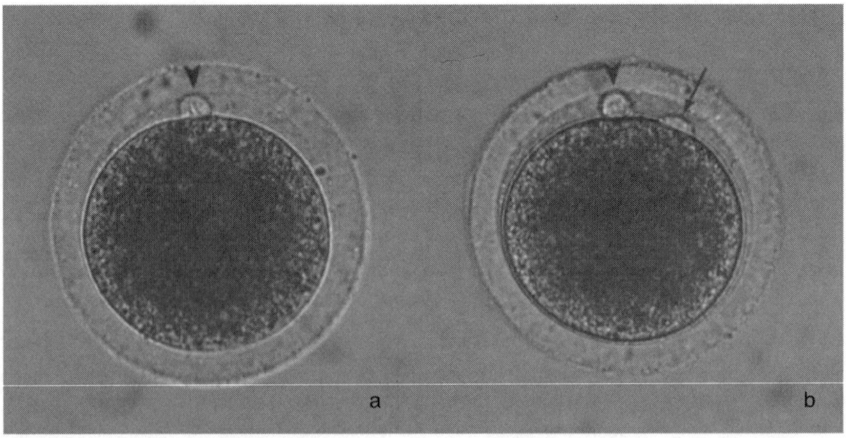

FIGURE 4.6 Bovine MII oocytes before (a) and after (b) treatment with 0.4 μg/ml demecolcine for 60 min. Arrow shows the membrane protrusion, and arrowhead indicates the first polar body.

G0/G1 phase of the cell cycle by serum starvation or contact inhibition. Single donor cells are inserted into the perivitelline space of recipient oocytes and both of them are fused with electric pulses (Figure 4.5). The fused oocytes are activated with cycloheximide or 6-dime thylaminopurine and cultured *in vitro* to the blastocyst stage and then nonsurgically transferred to synchronized recipients.

The potential of nuclear-transferred bovine oocytes receiving somatic cells to develop into blastocysts is 20 to 60%, but not largely different from that of *in vitro* fertilized oocytes. After transfer to recipients, the proportion of blastocysts that developed into young ranges from 5 to 80%, but is usually 10 to 20%.

4.4.2.1 Cell Cycle Stage

Although reprogramming of embryonic nuclei occurs in activated and nonactivated MII oocytes [33], reprogramming of somatic cell nuclei differs [58]. As shown in Figure 4.7, enucleated MII oocytes receiving cultured bovine cumulus cells developed into blastocysts irrespective of the cell cycle stage of donor cells [58]. When enucleated oocytes activated 6 h before were used for recipient cytoplasm, none of nuclear-transferred oocytes receiving cumulus cells at any cell cycle stage developed beyond the eight-cell stage. Because embryonic genome activation of bovine embryos occurs between the 8- to 16-cell stage [59], new transcription in the nuclear-transferred oocytes apparently did not occur. It is possible that MII oocytes reprogram somatic and embryonic cell nuclei but activated oocytes reprogram only embryonic cell nuclei.

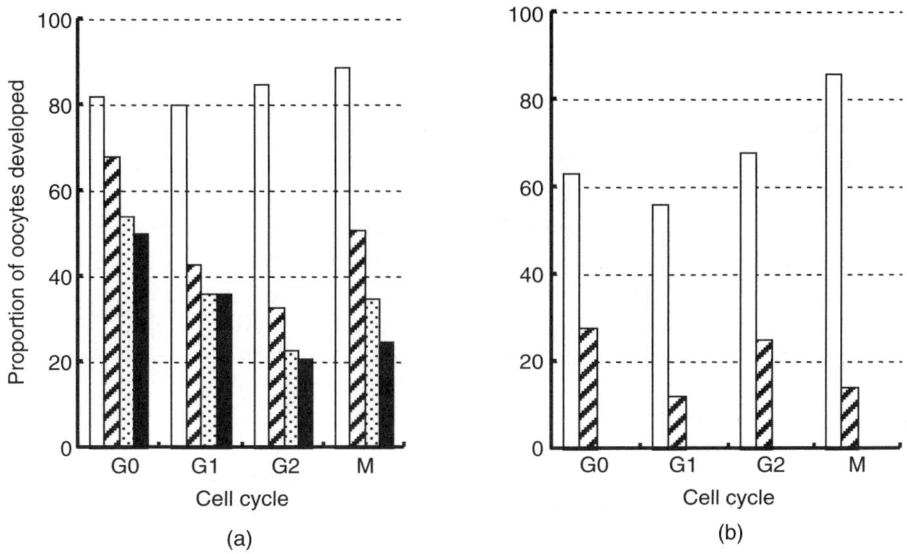

FIGURE 4.7 *In vitro* developmental potential of nuclear-transferred bovine oocytes at the M phase (a) or S phase (b) receiving adult somatic cells at different cell cycle stages.

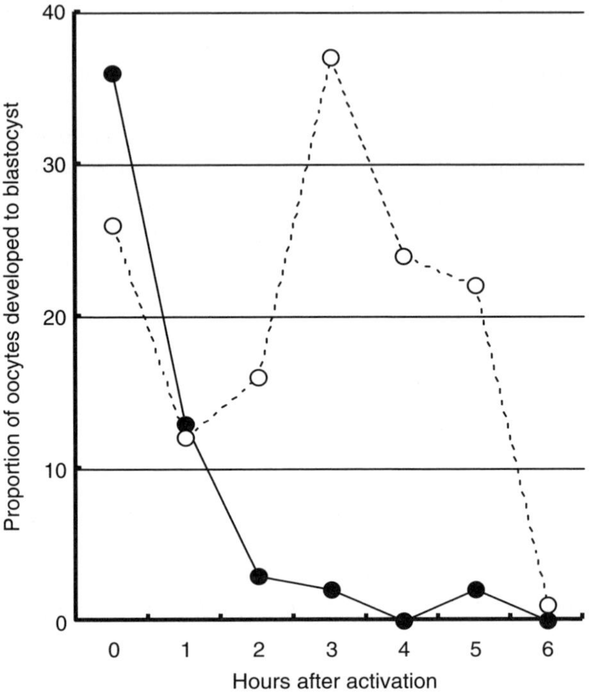

FIGURE 4.8 *In vitro* developmental potential of nuclear-transferred bovine oocytes at the G0/G1 phase (●–●) and M phase (○–○).

The point at which the potential of oocytes to reprogram somatic cell nuclei disappears after parthenogenetic activation was recently examined [36]. The potential of nuclear-transferred oocytes to develop into blastocysts is not different when cultured cumulus cells at the M phase are fused with oocytes activated 1 to 5 h before (Figure 4.8). The potential of oocytes activated 1 h before receiving somatic cells at the G0/G1 phase, however, decreases significantly. The decline is not due to a lack of reprogramming factors in oocytes, but rather to low MPF activity. It was demonstrated that reprogramming of bovine somatic cell nuclei is not directly regulated by mitogen-activating protein kinase activity, similar to MPF [36]. The reprogramming factor in oocytes, however, is not yet identified.

4.4.2.2 Donor Cell Type

Somatic cells from various tissues of male and female fetuses, newborns, and adults have been cultured, passaged, and used for bovine nuclear transfer:

- Cumulus (granulosa)
- Oviduct
- Uterus
- Skin
- Ear

- Heart
- Liver
- Kidney
- Muscle
- Lung
- Mammary gland
- Testis
- Epididymis
- Gut
- Tongue
- Bone marrow

The *in vitro* potential of enucleated bovine oocytes receiving somatic cells from various tissues to develop into blastocysts by using the same nuclear transfer procedures was compared [55]. The percentage of blastocysts that developed from enucleated oocytes receiving donor cells from various tissues was not much different among donor cells; also, no difference was found in the percentage of blastocysts that developed from enucleated oocytes receiving adult, newborn, or fetal cells, or between female and male cells (Table 4.9).

So far, cloned calves have been produced after nuclear transfer of somatic cells from many tissues, but the most suitable donor cells for cloning are not clear because the developmental potential of nuclear-transferred oocytes receiving somatic cells obtained from different tissues of a single animal has not been compared. Several studies have demonstrated that cell lines from cumulus cells (granulosa cells) are suitable donor cells for bovine cloning [54,55,60]. Different developmental potentials of nuclear-transferred oocytes receiving cell lines from the same origin have also been observed [55]. Although one live calf was obtained after nuclear transfer of pluripotent mesenchymal stem cells derived from male adult bone marrow [61], it is not clear whether somatic stem cells have a better chance to develop into live calves than differentiated somatic cells do.

TABLE 4.9
The Potential of Nuclear-Transferred Bovine Oocytes Receiving Somatic Cells from Different Origins to Develop into Blastocysts *In Vitro*

	Origin		
	Adult	Newborn	Fetus
Female	30–52	28–44	223–47
Male	33–53	25–47	37–45

TABLE 4.10
Frequency of Gestation Losses from Somatic Cell Cloned Bovine Embryos

	No. of Recipients (%)			No. of Calves (%)		
Reference	Transferred	Pregnant	Calved	Born	Born Alive	Remain Alive
54, 55	139	55 (40)	28 (20)	32	25 (78)	18 (56)
62	173	97 (56)	15 (9)	15	15 (100)	11 (73)
64	247	110 (45)	30 (12)	30	24 (80)	24 (80)
65	2170	535 (25)	103 (5)	117	106 (91)	82 (70)

4.4.3 Health Consequences

4.4.3.1 Pregnancy Loss

As shown in Table 4.10, the initial pregnancy rates after transfer of somatic cell-cloned bovine embryos are relatively high, but more than 50% of pregnant recipients abort before parturition. Pregnancy losses after day 90 of gestation are rare in bovine implanted by artificial insemination or embryo transfer of blastocysts fertilized *in vivo* and *in vitro*. Heyman et al. [62] reported that the incidence of fetal loss between day 90 and calving was 44% for adult somatic cell clones and 33% for fetal somatic cell clones, compared with 4.3% after embryonic cell cloning. In one study [55] in which blastocysts that originated from various somatic cells were transferred, the pregnancy rate gradually declined and abortions occurred even after day 150 of gestation (Figure 4.9). The main reasons for the high incidence of abortion might be due to the placental abnormalities in early and late-term cloned bovine fetuses [60,63,64].

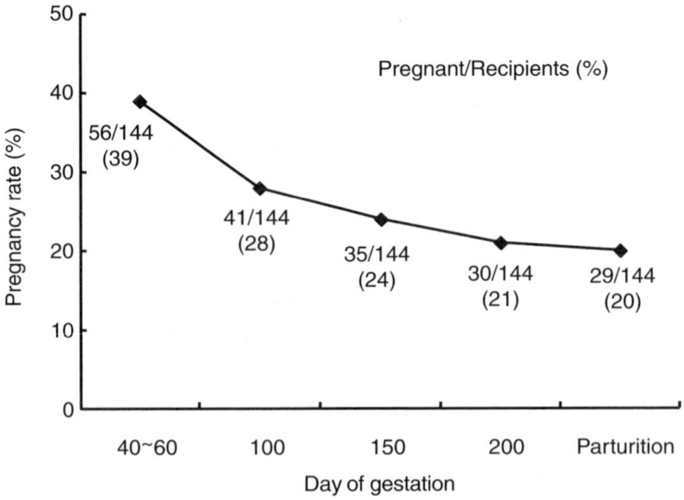

FIGURE 4.9 *In vivo* development after transfer of bovine cloned embryos.

4.4.3.2 Perinatal Death of Cloned Calves and Observed Abnormalities

As shown in Table 4.10, 56 to 80% of somatic cell clones were alive for at least for 1 week to 1 month after parturition [54,55,62,64,65], but the rest died for various reasons (Table 4.11). Since the first somatic cell-cloned calves were born on July 5, 1998, in Japan [54], 367 calves have been produced following nuclear transfer of somatic cells from various origins; however, 174 (47%) have died for various reasons (Table 4.8). The proportions of peri-and postnatal death of somatic cell clones were higher than those of embryonic cell clones (18%, Table 4.8).

Chavatte-Palmer et al. [66] reported that apparently normal surviving cloned calves cannot be considered physiologically normal animals until at least 50 days of age. Cloned neonates had high body temperature, high plasma leptin concentrations, and low T4 levels compared with artificially inseminated controls. These parameters returned to the same levels as those in controls 50 days after birth. Lanza et al. [67] reported that physical examination of surviving somatic cell clones was normal. This physiologic status is in accordance with the impression that, in general, most cloned neonates require intensive care such as oxygen inhalation, nitric oxide administration, body temperature maintenance, colostrum administration, etc. [68]. Abnormalities have also been observed in embryonic stem cell-cloned mice [69,70] and somatic cell-cloned sheep [71], goats [72], and pigs [73]. The body weight of cloned calves is frequently heavy, but large offspring syndrome is not a specific characteristic of the clones [74]. Such excessive weight in somatic cell clones, however, has not been observed in pigs [56,57].

The reasons for the high incidence of such abnormalities in neonates are not clear, but some might be due to the hydrallatois caused by placental abnormalities

TABLE 4.11
Abnormalities Observed in Cloned Calves

Pneumonia	Anasarca
Cardiopulmonary anomalies	Enlarged placentomes
Ventricular septal defects	Acidosis
Omphalocele	Enlarged right ventricle
Ascites	Pulmonary surfactant deficiency
Cardiac enlargement	Pulmonary hypertension
Liver steatosis	Meconium aspiration
Hypothermia	Enlarged umbilicus
Hypoxemia	Septicemia
Hypoglycemia	Immunodeficiency
Hypercarbia	Diabetes
Pulmonary hypertension	Idiopathic hyperthermia
Allantoic edema	Gastritis/enteritis
Hydrallantois	Collisepticemia
Hydramnios	Morphologic abnormalities
Hyperthermia	Large offspring syndrome

with a small number of placentomes [60,64,75]. Sudden death with no premonitory signs has been observed in some clones even at 5 to 6 months old [76]. In addition to abnormalities observed in cloned calves, most pregnant recipients had no or only minor symptoms of parturition, such as labor pain and mammary development [60,68,71]. Thus, delivery of the neonates is usually performed by cesarean section.

4.4.3.3 Postnatal Development and Reproductive Ability

Most somatic cell-cloned calves that survive 1 month after parturition will grow normally. Although the growth rates of cloned calves have not been reported, obesity in cloned mice [77] has been observed. Figure 4.10 shows the change in body weight of two bulls obtained from adult somatic cells of an elite bull from parturition to 30 months of age. The body weights of both clones are within the upper and lower limits of bulls of the same breed and age.

Several reports have demonstrated that somatic-cell cloned male and female calves have normal reproductive ability [67,76,78]. In one study, all 13 calves — 11 females and 2 males — that grew to adulthood had normal fertility [76]. The first four surviving calves, obtained from cultured cumulus cells and oviductal cells from July 5 to August 19, 1998 [54], produced normal calves on July 10, 2000; September 17, 2000; March 26, 2001; and April 23, 2001, after artificial insemination.

The first somatic cell-cloned mammals had shorter telomere lengths [79], but later studies demonstrated longer [80] lengths equivalent [81,82] to those of control

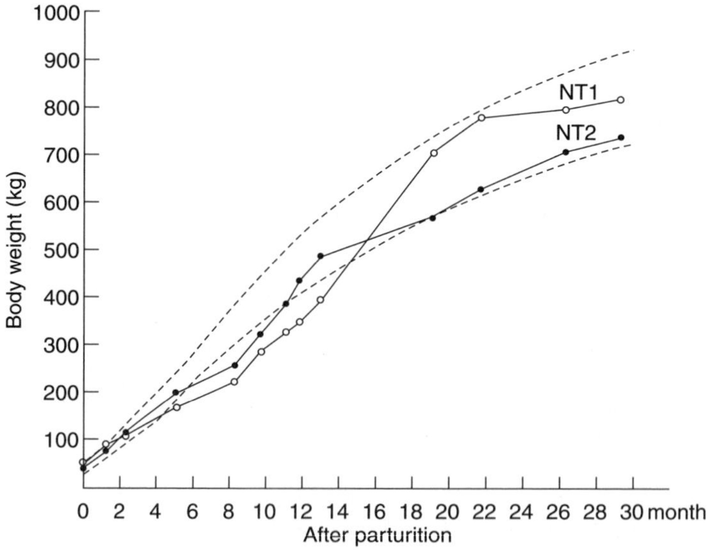

FIGURE 4.10 Body weights of somatic cell cloned bulls. NT1 and NT2 were cloned bulls obtained after nuclear transfer of fibroblast cells from an adult Japanese brown cattle. Dotted lines show upper and lower limits in control bulls.

animals, different among tissues examined [55], or different among donor cells [83]. The life span of cloned mice from somatic cells is significantly shorter than that of controls [84], but it is not known if this is the case in cloned calves.

4.5 SUMMARY

Various new technologies have been developed to increase the efficiency of embryo transfer in farm animals. Of these technologies, nuclear transplantation with somatic cells is the most efficient technique because copies of elite bovine can be produced in a short time. The cloning technology has some problems, however, especially with somatic cells, such as a high incidence of peri- and postnatal death and morphologic abnormalities of calves. More reliable cloning techniques need to be developed before broad application in animal husbandry.

ACKNOWLEDGMENT

This work was supported by grants from the Program for Promotion of Basic Research Activities for Innovative Biosciences (PROBRAIN), Ministry of Education, Science, and Culture (13308054, 14034259, 15039233), Ito Kinen Foundation, and Nakajima International Interchange Foundation.

REFERENCES

1. Willadsen, S.M., A method for culture of micromanipulated sheep embryos and its use to produce monozygotic twins, *Nature*, 277, 298, 1979.
2. Willadsen, S.M. and Polge, C., Attempts to produce monozygotic quadruplets in cattle by blastomere separation, *Vet. Rec.*, 108, 211, 1981.
3. Ozil, J.P., Heyman, Y., and Renard, J.P., Production of monozygotic twins by micromanipulation and cervical transfer in the cow, *Vet. Rec.*, 110, 126, 1982.
4. Kato, Y. and Tsunoda, Y., *Encyclopedia of Dairy Sciences*, Vol. 4, 3rd ed., Academic Press, New York, 1181, 2002.
5. Matsumoto, K. et al., Production of identical twins by separating two-cell rat embryos, *Gamete Res.*, 22, 257, 1989.
6. Tsunoda, Y., Yasui, T. and Sugie, T., Production of monozygotic twins following transfer of separated half embryos in the goat, *Jpn. J. Anim. Sci.*, 55, 643, 1984.
7. Allen, W.R. and Pashen, R.L., Production of monozygotic (identical) horse twins by embryo micromanipulation, *J. Reprod. Fertil.*, 71, 607, 1984.
8. Yang, X.Z. and Foote, R.H., Production of identical twin rabbits by micromanipulation of embryos, *Biol. Reprod.*, 37, 1007, 1987.
9. Willadsen, S.M. and Godke, R.A., A simple procedure for the production of identical sheep twins, *Vet. Rec.*, 114, 240, 1984.
10. Gatica, R. et al., Micromanipulation of sheep morulae to produce monozygotic twins, *Theriogenology*, 21, 555, 1984.
11. Tsunoda, Y. et al., Production of monozygotic twins following the transfer of bisected embryos in the goats, *Theriogenology*, 24, 337, 1985.
12. Nagashima, H. et al., Production of normal piglets from microsurgically split morulae and blastocysts, *Theriogenology*, 29, 485, 1988.

13. Reichelt, B. and Niemann, H., Generation of identical twin piglets following bisection of embryos at the morula and blastocyst stage, *J. Reprod. Fertil.*, 100, 163, 1994.
14. Lambeth, V.A. et al., Microsurgery on bovine embryos at the morula stage to produce monozygotic twin calves, *Theriogenology*, 20, 85, 1983.
15. Arave, C.W. et al., Factors affecting survivability of transferred whole and demi-embryos in a commercial dairy herd, *Theriogenology*, 28, 373, 1987.
16. McEvoy, T.G. and Sreenan, J.M., Effect of embryo quality and stage of development on the survival of zona pellucida-free cattle demi-embryos, *Theriogenology*, 33, 1245, 1990.
17. Chan, A.W.S. et al., Clonal propagation of primate offspring by embryo splitting, *Science*, 287, 317, 2000.
18. Voelkel, S.A. et al., Multiple embryo-transplant offspring produced from quartering a bovine embryo at the morula stage, *Vet. Rec.*, 117, 528, 1985.
19. Willadsen, S.M., The developmental capacity of blastomeres from four- and eight-cell sheep embryos, *J. Embryol. Exp. Morph.*, 65, 165, 1981.
20. Willadsen, S.M. and Fehilly, C.B., *Fertilization of Human Egg in Vitro*, Springer–Verlag, Berlin, 1983, 353.
21. Kelly, S.J., Studies of the developmental potential of four- and eight-cell stage mouse blastomeres, *J. Exp. Zool.*, 200, 365, 1977.
22. Tsunoda, Y. et al., Developmental of one or two blastomeres from eight-cell mouse embryos to term in the presence of parthenogenetic eggs, *Theriogenology*, 28, 615, 1987.
23. Leibo, S.P. and Rall, W.F., Increase in production of pregnancies by bisection of bovine embryos, *Theriogenology*, 27, 245, 1987.
24. Kippax, I.S., Christie, W.B., and Rowan, T.G., Effect of method of splitting, stage of development and presence or absence of zona pellucida on fetal survival in commercial bovine embryo transfer of bisected embryos, *Theriogenology*, 35, 25, 1991.
25. Gray, K.R., Bondioli, K.R., and Betts, C.L., The commercial application of embryo splitting in beef cattle, *Theriogenology*, 35, 37, 1991.
26. Tsunoda, Y. et al., Beneficial effect of agar for the frozen storage of bisected embryos, *Theriogenology*, 28, 317, 1987.
27. Tsunoda, Y. and McLaren, A., Effect of various procedures on the viability of mouse embryos containing half the normal number of blastomeres, *J. Reprod. Fertil.*, 69, 315, 1983.
28. McGrath, J. and Solter, D., Nuclear transplantation in the mouse embryo by microsurgery and cell fusion, *Science*, 220, 1300, 1983.
29. McGrath, J. and Solter, D., Inability of mouse blastomere nuclei transferred enucleated zygotes to support development *in vitro*, *Science*, 226, 1317, 1984.
30. Tsunoda, Y. et al., Full-term development of mouse blastomere nuclei transplanted into enucleated two-cell embryos, *J. Exp. Zool.*, 242, 147, 1987.
31. Willadsen, S.M., Nuclear transplantation in sheep embryos, *Nature*, 320, 63, 1986.
32. Prather, R.S. et al., Nuclear transplantation in the bovine embryo: assessment of donor nuclei and recipient oocyte, *Biol. Reprod.*, 37, 859, 1987.
33. Campbell, K.H.S. et al., Cell cycle coordination in embryo cloning by nuclear transfer, *Rev. Reprod.*, 1, 40, 1996.
34. Yang, X. et al., Synergistic effect of ethanol and cycloheximide on activation of freshly matured bovine oocytes, *Theriogenology*, 41, 395, 1994.
35. Kato, Y. and Tsunoda, Y., Synchronous division of mouse two-cell embryos with nocodazole *in vitro*, *J. Reprod. Fertil.*, 95, 39, 1992.

36. Tani, T., Kato, Y., and Tsunoda, Y., Reprogramming of bovine somatic cell nuclei is not directly regulated by maturation promoting factor or mitogen-activated protein kinase activity, *Biol. Reprod.*, 69, 1890, 2003.
37. Collas, P. and Barnes, F.L., Nuclear transplantation by microinjection of inner cell mass and granulose cell nuclei, *Mol. Reprod. Dev.*, 38, 264, 1994.
38. Keefer, C.L., Stice, S.L., and Matthews, D.L., Bovine inner cell mass cells as donor nuclei in the production of nuclear transfer embryos and calves, *Biol. Reprod.*, 50, 935, 1994.
39. Willadsen, S.M. et al., The viability of late morulae and blastocysts produced by nuclear transplantation in cattle, *Theriogenology*, 35, 161, 1991.
40. Sims, M. and First, N.L., Production of calves by transfer of nuclei from cultured inner cell mass cells, *Proc. Natl. Acad. Sci., USA*, 90, 6143, 1993.
41. Brink, M.F., Bishop, M.D., and Pieper, F.R., Developing efficient strategies for the generation of transgenic cattle which produce biopharmaceuticals milk, *Theriogenology*, 53, 139, 2000.
42. Forsberg, E.J. et al., Cloning nontransgenic and transgenic cattle, *Theriogenology*, 55, 269, 2001 (abstr.).
43. Ohkoshi, K. et al., Effects of the age of donor embryos on the developmental ability of bovine nuclear transferred eggs *in vitro*, *J. Reprod. Dev.*, 43, 261, 1997.
44. Westhusin, M.E., Pryor, J.H., and Bondioli, K.R., Nuclear transfer in the bovine embryo: a comparison of 5-day, 6-day, frozen–thawed, and nuclear transfer donor embryos, *Mol. Reprod. Dev.*, 28, 119, 1991.
45. Tsunoda, Y. et al., Nuclear transplantation of male primordial germ cells in the mouse, *Development*, 107, 407, 1989.
46. Kato, Y. et al., Developmental potential of mouse primordial germ cells, *Development*, 126, 1823, 1999.
47. Zakhartchenko, V. et al., Potential of fetal germ cells for nuclear transfer in cattle, *Mol. Reprod. Dev.*, 52, 421, 1999.
48. Willadsen, S.M., Cloning of sheep and cow embryos, *Genome*, 31, 956, 1989.
49. Stice, S.L. and Keefer, C.L., Multiple generational bovine embryo cloning, *Biol. Reprod.*, 48, 715, 1993.
50. Ectors, F.J. et al., Viability of cloned bovine embryos after one or two cycles of nuclear transfer and *in vitro* culture, *Theriogenology*, 44, 925, 1995.
51. Takano, H. et al., Cloning of bovine embryos by multiple nuclear transfer, *Theriogenology*, 47, 1365, 1997.
52. Wilmut, I. et al., Viable offspring derived from fetal and adult mammalian cells, *Nature*, 385, 810, 1997.
53. Tsunoda, Y. and Kato, Y., Recent progress and problems in animal cloning, *Differentiation*, 69, 158, 2002.
54. Kato, Y. et al., Eight calves cloned from somatic cells of a single adult, *Science*, 282, 2095, 1998.
55. Kato, Y., Tani, T., and Tsunoda, Y., Cloning of calves from various somatic cell types of male and female adult, newborn and fetal cows, *J. Reprod. Fertil.*, 120, 231, 2000.
56. Yin, X.J. et al., Production of cloned pigs from adult somatic cells by chemically assisted removal of maternal chromosomes, *Biol. Reprod.*, 67, 442, 2002.
57. Kawakami, M. et al., Effect of demecolcine and nocodazole on the efficiency of chemically-assisted removal of chromosomes and the developmental potential of nuclear transferred porcine oocytes, *Cloning Stem Cells*, 5, 379, 2003.

58. Tani, T., Kato, Y., and Tsunoda, Y., Direct exposure of chromosomes to nonactivated ovum cytoplasm is effective for bovine somatic cell nucleus reprogramming, *Biol. Reprod.*, 64, 324, 2001.
59. Camous, S., Kopecny, V., and Flechon, J.E., Autoradiographic detection of the earliest stage of [3H]-uridine incorporation into the cow embryo, *Biol. Cell*, 58, 195, 1986.
60. Wells, D.N., Misica, P.M., and Tervit, H.R., Production of cloned calves following nuclear transfer with cultured adult mural granulose cells, *Biol. Reprod.*, 60, 996, 1999.
61. Kato, Y. et al., Nuclear transfer of adult bone marrow mesenchymal stem cells: developmental totipotency of tissue-specific stem cells from an adult mammal. *Biol, Reprod.*, 70, 415, 2004.
62. Heyman, Y. et al., Frequency and occurrence of late-gestation losses from cattle cloned embryos, *Biol. Reprod.*, 66, 6, 2002.
63. Stice, S.L. et al., Pluripotent bovine embryonic cell lines direct embryonic development following nuclear transfer, *Biol. Reprod.*, 54, 100, 1996.
64. Hill, J.R. et al., Evidence for placental abnormality as the major cause of mortality in first-trimester somatic cell cloned bovine fetuses, *Biol. Reprod.*, 63, 1787, 2000.
65. Pace, M.M. et al., Ontogeny of cloned cattle to lactation, *Biol. Reprod.*, 67, 334, 2002.
66. Chavatte-Palmer, P. et al., Clinical, hormonal, and hematologic characteristics of bovine calves derived from nuclei from somatic cells, *Biol. Reprod.*, 66, 1596, 2002.
67. Lanza, R.P. et al., Cloned cattle can be healthy and normal, *Science*, 294, 1893, 2001.
68. Hill, J.R. and Chavatte-Palmer, P., *Principles of Cloning*, Academic Press, Amsterdam, 247, 2002.
69. Wakayama, T. et al., Mice cloned from embryonic stem cells, *Proc. Natl. Acad. Sci. USA*, 96, 14984, 1999.
70. Amano, T., Kato, Y., and Tsunoda, Y., Full-term development of enucleated mouse oocytes fused with embryonic stem cells from different cell lines, *Reproduction*, 121, 729, 2001.
71. Schnieke, A.E. et al., Human factor IX transgenic sheep produced by transfer of nuclei from transfected fetal fibroblasts, *Science*, 278, 2130, 1997.
72. Zou, X. et al., Production of cloned goats from enucleated oocytes injected with cumulus cell nuclei or fused with cumulus cells, *Cloning Stem Cells*, 3, 31, 2001.
73. Carter, D.B. et al., Phenotyping of transgenic cloned piglets, *Cloning Stem Cells*, 4, 131, 2002.
74. Yong, L.E., Sinclair, K.D., and Wilmut, I., Large offspring syndrome in cattle and sheep, *Rev. Reprod.*, 3, 155, 1998.
75. DeSousa, P.A., Evaluation of gestational deficiencies in cloned sheep fetuses and placentae, *Biol. Reprod.*, 65, 23, 2001.
76. Tsunoda, Y. and Kato, Y., Donor cell type and cloning efficiency in mammals. In Cibelli, J., Lanza, R.P., Campbell, K.H.S., and West, W.D. (Eds.), *Principles of Cloning*, New York: Academic Press, 267, 2002.
77. Tamashiro, K.L.K. et al., Postnatal growth and behavioral development of mice cloned from adult cumulus cells, *Biol. Reprod.*, 63, 328, 2000.
78. Enright, B.P. et al. Reproductive characteristics of cloned heifers derived from adult somatic cells, *Biol. Reprod.*, 66, 291, 2002.
79. Shiels, P.G. et al., Analysis of telomere lengths in cloned sheep, *Nature*, 399, 316, 1999.
80. Lanza, R.P. et al., Extension of cell life-span and telomere length in animals cloned from senescent somatic cells, *Science*, 288, 665, 2000.

81. Tian, X.C., Xu, J., and Yang, X., Normal telomere lengths found in cloned cattle, *Nat. Genet.*, 26, 272, 2000.
82. Betts, D.H. et al., Reprogramming of telomerase activity and rebuilding of telomere length in cloned cattle, *Proc. Natl. Acad. Sci. USA*, 98, 1077, 2001.
83. Miyashita, N. et al., Remarkable differences in telomere lengths among cloned cattle derived from different cell types, *Biol. Reprod.*, 66, 1649, 2002.
84. Oganuki, N. et al., Early death of mice cloned from somatic cells, *Nat. Genet.*, 30, 253, 2002.

5 Cloning the Equine

Kenneth L. White, Gordon L. Woods, Dirk K. Vanderwall, Guanpeng Li, and Thomas D. Bunch

CONTENTS

5.1 INTRODUCTION

Nuclear transfer has been used to produce mammalian offspring in sheep, cattle, pigs, mice, rabbits, goats, cats, deer, rat, mules, and the horse. Live animals have been produced from somatic (nonreproductive) cells from adult and fetal cell lines. Recently, the authors' laboratory reported the first successful births of an equine specie — three identical cloned mules produced from the transfer of fetal somatic cell nuclei into enucleated horse oocytes and carried full term in the reproductive tract of a female horse (mare) [1]. All of these animals developed normally and the births were unassisted; each animal is developmentally normal and healthy based on several physiological parameters monitored since birth. Subsequent to the successful birth of the first equine, Galli et al. [2] reported the birth of a single horse foal produced following the transfer of adult somatic cells into an enucleated horse oocyte. This animal was reported to be normal and healthy.

5.2 UNIQUE OBSTACLES TO CLONING THE FIRST EQUINE

Success in the equine was extremely remarkable considering the existing inability to produce efficient numbers of *in vitro* produced oocytes and embryos with the ability to develop to term [3]. Both of these assisted reproductive parameters are well established in each of the species that have produced nuclear transfer offspring and aspects of these processes contribute to the efficiency of the cloning success. Initial and, in some species, continued success of the nuclear transfer process depends on generating large numbers of embryos for transfer. The large number of embryos is requisite for success because of the extremely high loss of pregnancies throughout gestation.

The authors' work in the mule and that of Galli et al. [2] in the horse used *in vitro* matured oocytes as the recipient cell for nuclear transfer. In the mule work, the oocytes were collected from live animals and further matured in culture for 16 h compared to the horse work, in which oocytes were collected from abattoir-derived ovaries and recovered oocytes were further cultured for 24 h. In addition, the single nuclear transfer horse foal resulted from the transfer of an adult somatic cell nucleus as compared to the mule foals that resulted from a fetal primary somatic cell line. A very unique aspect of the horse foal is that its dam is, in fact, its genetically identical twin. This clearly indicates, at least in the equine, that recognition of fetal antigens is not necessary to allow pregnancy to proceed to term.

Of further significance is that the first equine produced used nuclear donor cells from a cell line of a mule, a sterile equine hybrid. Mules are the result of breeding a donkey with a horse and produce an animal with 63 chromosomes (donkey = 62 and horse = 64). These animals cannot reproduce naturally under any combination of breeding crosses [4]. Others have suggested the potential for cloning to allow the multiplication of genetically unique or valuable animals [5], to produce animals with altered genetic composition [6], and to facilitate the rescue of endangered species [7]. This is the first example of applying cloning as a tool to reproduce a species otherwise incapable of reproduction by natural means.

5.3 NUCLEAR DONOR CELLS

The primary donor cell line used to produce the identical mule foals was established from a 45-day old fetus recovered from a quarter horse type of mare by transcervical uterine lavage. Fetal tissues were aseptically collected and then dispersed; a primary culture of fibroblast cells was established. The cells were allowed to multiply for approximately 5 weeks, then stored frozen in liquid nitrogen until immediately prior to use as nuclear donor cells. The primary donor cell line comprised fibroblast cells as determined by labeling with vimentin and cytokeratine, a characteristic test of fibroblast cell linage.

5.4 EQUINE EXTRACELLULAR AND INTRACELLULAR CALCIUM

A critical component of the cloning process in any species is the efficient activation of the single-cell nuclear transfer zygote following nuclear transplantation. Unlike other cell types, oocytes (eggs) are maintained in an immature state until immediately prior to release from the ovary. The oocyte matures to a point within meiosis, metaphase II stage, and arrests at this developmental point until fertilized. This "activates" the oocyte to complete meiosis and begin zygotic development. Little is known regarding the intracellular mechanisms associated with oocyte activation in the equine. However, in all other species studied, intracellular calcium transients are an absolute requirement for initiation of subsequent development [8].

Ongoing, independent research evaluating the differences between equine and human cells relative to growth characteristics indicated that average (n = 55) intracellular calcium levels in equine erythrocytes were 2.5-fold lower than average (n = 55) human erythrocyte levels — even though extracellular calcium levels were 1.5-fold higher in the equine (unpublished data, Woods et al.). Based on this information, calcium levels were evaluated in follicular fluid obtained from the ovaries of eight mares and compared to comparable samples obtained from the ovaries of seven cows. These results indicated that equine follicular fluid contained 38.6-fold higher calcium levels (average 109,965 ng/g vs. 2849 ng/g, respectively) than similarly staged bovine follicular fluid (unpublished data, Woods et al.).

5.5 EQUINE OOCYTE CALCIUM REGULATION

Considering these results, initial studies were carried out to evaluate the ability of equine oocytes to initiate an intracellular calcium transient when exposed to ionomycin — a known and commonly utilized intracellular calcium agonist. The ability of the oocyte to generate a calcium transient of sufficient magnitude to induce oocyte activation was requisite to any subsequent development.

Preovulatory oocytes were collected from ≥45-mm follicles by transvaginal ultrasound-guided aspiration. Five horse oocytes were recovered from five different mares cultured for 14 h and used to evaluate intracellular calcium response. Using procedures common to the authors' laboratory, these equine oocytes, along with bovine *in vitro* matured oocytes used for comparative purposes, were evaluated for intracellular calcium response following treatment with ionomycin [9]. The intracellular calcium profile observed after treatment with ionomycin in equine oocytes was markedly reduced compared to that obtained with bovine oocytes under the same conditions.

Evaluated together, these results indicated that the equine has unique calcium regulatory mechanisms — particularly considering the large disparity between the extracellular calcium concentrations observed in the equine and the correspondingly low intracellular levels detected. Furthermore, combined with the equine oocyte intracellular calcium response induced by ionomycin, these results provided the impetus for increasing the calcium levels in media associated with the equine oocytes and nuclear transfer embryos during *in vitro* culture and manipulation. This modification to increase the calcium content (sixfold) of media exposed to the oocytes and subsequent nuclear transfer zygotes was critical to the success of the cloning procedure and led to the production of the three identical mule foals. Following this modification, the 2-week pregnancy rate following transfer of equine nuclear transfer embryos increased over sevenfold compared to all other treatment groups.

5.6 NUCLEAR TRANSFER OF EQUINE EMBRYOS

Oocytes to be used for nuclear transfer were collected by transvaginal ultrasound-guided pick-up from quarter horse types of donor mares during the preovulatory period from ≥45-mm Graafian follicles. Oocytes were immediately recovered from flush medium, graded based on morphology, and placed into culture for an additional 12 to 16 h prior to nuclear transfer. Oocytes were classified into one of three categories:

- Expanded (with large, loose cumulus complexes)
- Compact (with tightly packed cumulus or coronal cells)
- Denuded (free of cumulus cells)

Regardless of classification, all oocytes were placed individually in maturation medium containing elevated calcium levels and cultured for 12 to 16 h. After culture, oocytes were denuded of cumulus cells. The first polar body and adjacent cytoplasm containing the metaphase plate were removed and a nuclear donor cell was placed in the perivitelline space closely associated to the enucleated oocyte. The resultant nuclear transfer couplets were placed into fusion medium within a fusion chamber and induced to fuse by electrical pulse.

Couplets that failed to fuse after the first fusion pulse were subjected to subsequent electrical pulses until fusion occurred or the enucleated oocyte lysed. The resultant nuclear transfer embryos were then placed in medium containing elevated (six times) calcium and activated by a brief exposure to ionomycin followed a 5-h treatment with cyclohexamide (Figure 5.1).

A total of 334 oocytes across all treatment groups were manipulated; 307 were successfully fused and 305 were transferred to recipient mares. Among these nuclear transfer zygotes, 6.9% resulted in 14-day pregnancies, 3.6% in 30-day pregnancies, and 1.6% in 45-day pregnancies. Of those nuclear transplant embryos maintained in standard calcium levels (1×), only 1.5% established 14-day pregnancies and none developed to 30-day pregnancy. Of the 113 nuclear transfer embryos cultured and handled in 3× media and activated in 6× medium were transferred to recipient mares, 12.4% resulted in 14-day pregnancy rates, 8.0% in 30-day pregnancy rates, and 4.4% in 45-day pregnancy rates; three live foals resulted.

The three foals born developed to full term and parturition proceeded naturally without assistance. All three foals were developmentally normal with respect to respiration, cardiovascular function, blood glucose levels, urinary output, and digestion at birth and continue to exhibit normal growth characteristics.

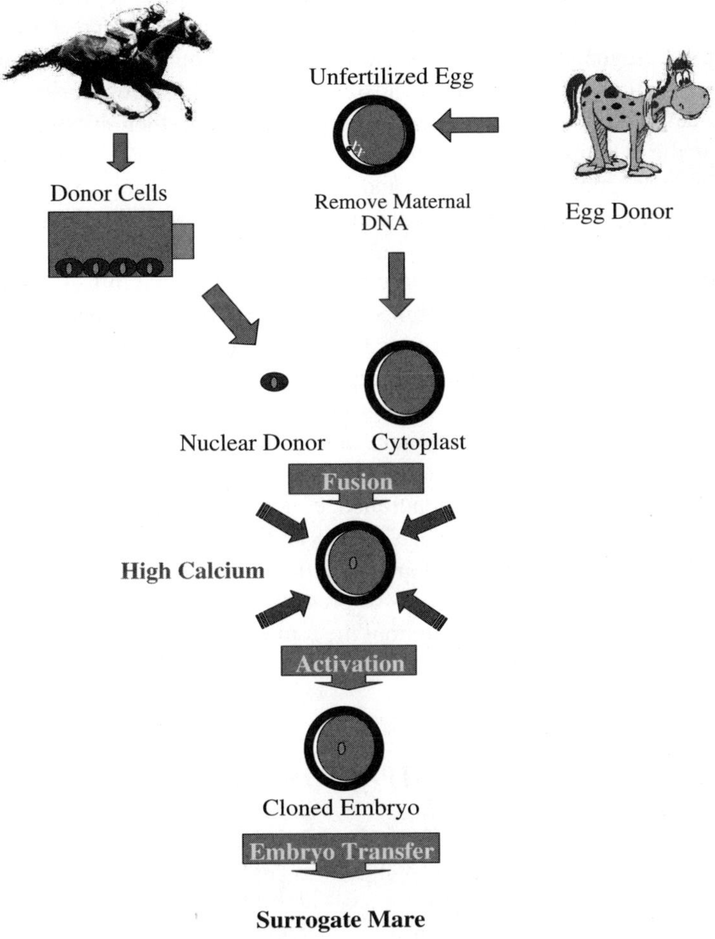

FIGURE 5.1 Diagrammatic illustration of the nuclear transfer process applied to the equine. Unfertilized eggs are recovered from the ovary of any grade mare, all maternal DNA is microsurgically removed and the resulting DNA-free egg is referred to as a cytoplast. The cytoplast is then combined with a single donor cell that contains a nucleus and the two cells are induced to fuse by electro-fusion. The nuclear transfer embryo is activated under a high calcium environment, which induces the initiation of development, and the embryo is then immediately transferred to a synchronized surrogate mare.

5.7 GESTATIONAL DEVELOPMENT OF CLONED MULES

The three quarter horse type surrogate mares carrying the cloned mule pregnancies were monitored throughout the gestation period. Mares were of various ages (8, 10, and 12 years) with unknown reproductive histories. Immediately following the completion of the nuclear transfer process, four one-cell cloned mule embryos were transferred surgically to the oviduct ipsilateral to the ovary of ovulation in each mare. Ovulation in surrogate mares had occurred during the 24 h preceding transfer.

An initial pregnancy determination was carried out by transrectal ultrasonography at 13 to 16 days of gestation (transfer day = day 0). Subsequent evaluations were conducted at 7- to 10-day intervals until day 60 and then every 2 to 4 weeks throughout the remainder of gestation. During the first 60-day evaluation period, pregnancies were monitored for size and location of embryonic vesicles, presence of an embryo within each vesicle, and embryonic heartbeat. The primary parameter evaluated during the 60- to 270-day mid-gestational evaluation period was fetal movement. The late gestational period was monitored for fetal movement and ultrasonographic assessment of the combined thickness of the uterus and placenta [10].

Of the cloned embryos transferred to these three surrogate mares, four developed into ultrasonographically detectable conceptuses. The growth profiles of each cloned conceptus were compared to published growth profiles of noncloned horse conceptuses that exhibited normal development during the same period as the evaluation. All four cloned conceptuses developed within normal size range throughout the evaluation period compared to noncloned conceptuses. Embryos were detected within each conceptus vesicle between day 21 and 29; however, a heartbeat was not detected in one of the twin pregnancies. This twin subsequently spontaneously eliminated between day 29 and day 36 of pregnancy. The three remaining embryos continued to develop normally throughout the remainder of gestation. All evaluations throughout gestation appeared to be within normal ranges for all parameters evaluated [10].

5.8 PARTURITION AND POSTPARTUM HEALTH OF CLONED MULES

As the mares approached their due dates, they were monitored for clinical signs indicative of preparation for parturition. All three mares exhibited the typical temporal pattern of change prior to parturition characteristic of noncloned equine pregnancies. When parturition appeared close (within 96 h), the mares were monitored by direct visual observation during the day, when they were turned out on pasture, and closed-circuit video during the night, when they were confined to a box stall.

The three mares spontaneously initiated and completed parturition without assistance on days 340, 346, and 346, respectively. All three foals were active and clinically healthy at birth, and all three mares spontaneously expelled the placenta within 90 min of parturition. Physical characteristics of each placenta were evaluated and compared to previously published data for Thoroughbred horses. One point of interest was the finding that the nonpregnant horn was longer than the pregnant horn in all three cloned placentas, whereas the pregnant horn is typically longer than the nonpregnant horn in equine placentas. The clinical significance (if any) of this finding is unknown [10].

Immediately after birth, all three foals were weighed and underwent a complete physical examination. In addition, they received routine postpartum care. All three stood unassisted within 30 min of birth. The three foals received 150 to 250 ml of a commercial equine IgG formulation via nasogastric tube within 4 h of birth. Within 36 h of birth, blood samples were collected from each foal for routine hematology and blood chemistry analyses. All three foals exhibited a mild to moderate thrombocytopenia, which resolved spontaneously and was not associated with any clinical signs. The underlying cause of the thrombocytopenia is unknown, but it may have been the result of absorption of alloantibodies in colostrum that recognized foal platelets. Clinically, the foals remained healthy throughout the neonatal period and continue to be healthy and vigorous at the time of this writing [10].

5.9 CONFIRMATION OF GENETIC ORIGIN

The confirmation of lineage of the nuclear transfer foal was by DNA testing for 12 horse microsatellite markers and sex-specific alleles of the amelogenin (AMEL) locus. These assays were performed with DNA extracted from cultured cells from the donor mule cell line and from blood and hair for the recipient mare and foal. These markers are used routinely for parentage verification and individual identification. Tests were independently performed at the Veterinary Genetics Laboratory, University of California at Davis. Results from these assays confirm that the three foals originated from the donor cell line and had no relation to their recipient mares [1].

5.10 THE ROLE OF CALCIUM IN EQUINE EMBRYO DEVELOPMENT

Apparently, elevated extracellular cellular calcium levels play a critical role during oocyte culture, handling, and activation that induces subsequent development of competent nuclear transfer zygotes. The success of these results also provides evidence of the potentially positive effect of elevated levels of extracellular calcium on equine embryonic/germ cells while they are maintained *in vitro*. The exposure of these cells to elevated calcium levels effectively resulted in a significant increase in the number of established pregnancies that subsequently developed to late gestation (>260 days) and finally to term.

These results may have further implications for implementing elevated extracellular calcium levels in conjunction with other assisted reproductive techniques, such as *in vitro* maturation, fertilization, culture, and sperm injection in the equine. The current success of these procedures in the equine is reduced compared to that with other species for unknown reasons [3]. The data reported here may have important implications for understanding fundamental differences in the cell biology of equine cells and their regulation of calcium that have important effects on subsequent developmental competence in this specie.

5.11 DEVELOPMENTALLY CRITICAL PERIOD

A critical period of early peri-implantation development in the equine may exist in which the majority of embryonic losses occur after transfer of cloned embryos. Of the total number of 14- to 30-day pregnancies established from all treatments (19), nine were ultrasonographically observed with clear, detectable fetal heartbeats. All fetuses with detectable heartbeats were from treatments with elevated calcium; eight of the nine were from the threefold/sixfold calcium group (one fetus was from the threefold activation treatment).

In addition, of the pregnancies with detectable heartbeats, in all pregnancies that subsequently failed, the heartbeats were lost prior to day 50 and the resulting vesicle disappeared within 20 days. This supports the hypothesis that critical events occur within the developing conceptus during 30 to 50 days of gestation; these result in a successful term pregnancy when appropriately carried out and failure when inadequate. It is also possible that critical interactions crucial to the survival of the developing fetus and pregnancy may occur between the developing fetal placenta and maternal endometrial tissue during this period.

A unique event that occurs during an equine pregnancy is the establishment of endometrial cups. These cups produce the hormone equine chorionic gonadotropin (eCG), which is believed to be responsible for maintenance of early pregnancy in the mare. The eCG is first detected in the serum of pregnant mares at 37 to 40 days, with peak levels reached at 60 to 80 days.

Endometrial cups are composed of tissue originating from the fetal placenta and maternal endometrial cells. Specialized tissue originating from the fetal placenta begins to appear at approximately 30 days of gestation and is referred to as chorionic girdle cells [11]. Specialized trophoblast cells originating from the chorionic girdle begin to migrate into the maternal endometrium at 36 to 38 days and become fully developed and functional by 45 to 50 days of gestation in pregnancies that develop to term [12]. The timing of these events coincides with the period of pregnancy when all cloned fetuses with heartbeats that failed to proceed to term were lost.

A problem with reprogramming equine somatic cell nuclei following nuclear transfer may be manifested as a placental problem, similar to the primary cause of pregnancy failure in other species in which successful births of nuclear transfer offspring have been previously reported [13,14]. The placental problem may be manifested as an inability to establish a competent presumptive population of chorionic girdle cells from the developing placenta; this leads to pregnancy failure. Further studies must be carried out to evaluate this possibility.

5.12 DEVELOPMENTAL COMPETENCE OF MULE CELLS

Results confirming the birth of cloned mules reported by the authors' laboratory provide evidence of developmental competence of mule somatic cells as nuclear donor cells. Mules are a sterile hybrid [4] resulting from the cross of a donkey with a horse. This developmental failure has been hypothesized to be associated with an incompatibility in chromosome number of the mule [4].

Mules have 63 chromosomes compared to 62 and 64 in the donkey and horse, respectively. Clearly, based on the birth of three identical mule foals, the mule nucleus has the endogenous factors required for directing development to a normal full-term offspring. Also, when maintained under the conditions of high extracellular calcium, the horse oocyte has the full potential to reprogram equine somatic cell nuclei; this results in the production of apparently healthy full-term offspring. The authors have reported that horse oocytes with 64 chromosomes are compatible with donkey sperm (fewer chromosomes) and mule nuclear transfer with 63 chromosomes. In contrast, donkey oocytes with 62 chromosomes are markedly less functional with stallion sperm (more chromosomes) [4]. In addition, using the elevated extracellular calcium treatment, attempts to use donkey oocytes as recipient cytoplasm have been made.

Results from these efforts yielded zero week-2 embryos when mule nuclei with 63 chromosomes were transferred to donkey oocytes ($n = 14$) with 62 chromosomes (unpublished data). This may support the hypothesis that critical, maternally imprinted genes are present on the horse complement of chromosomes that contribute to and are necessary for survival of the resulting mule. If this hypothesis is true, the birth of mule nuclear transfer offspring indicates a process of reprogramming that allows appropriate function of an imprinted gene.

5.13 REPRODUCTION OF INFERTILE OR STERILE ANIMALS

The ability to produce offspring from animals otherwise incapable of reproduction has important implications in the equine. Routinely, male animals are electively castrated to produce animals (geldings) that are more easily managed and handled. These animals are then much more conveniently and safely used in various competitions such as cutting, reining, polo, roping, and racing. An example of this situation was the success of the gelding, Funny Cide, that won two of the three legs of the Triple Crown in 2003. However, regardless of the potential superiority of the genetics of this animal without the application of nuclear transfer, he will never reproduce and pass his genes on to future generations.

Often the decision to geld an animal is made many months or years prior to the animal's beginning competition. Most animals will not begin competition prior to their third year of age. Unfortunately, although these animals are proven valuable competitors with outstanding gene-based traits, this is identified subsequent to castration. In these cases, nuclear transfer provides the only means by which these animals can contribute to the gene pool. A proven sterile animal can provide a minor tissue sample with no risk to its health and well-being, thus allowing the establishment of somatic cell lines that can be used as donor cells to produce a genetically identical animal capable of reproduction. Through this process, the animal's reproductive potential can be recovered and realized.

The ability to recover the genetics of animals that die prematurely or die prior to sufficient impact of valuable genomics is an important consideration in the horse. What is the economic impact of producing another genetically identical copy of Secretariat? What if the reproductive difficulties experienced by this great horse were environmental in nature rather than genetic? What if the conception rate of his clone were closer to 80% rather than less than 40%? Other great animals come to mind that would have a tremendous financial impact on the industry as well as a potentially significant impact on overall performance characteristics in the breed if their genetics were still accessible.

Obviously, in the case of the duplication of a mule (a sterile hybrid), nuclear transfer provides the only means for replication of genetic traits of specific animals. This may also be extended to examples of endangered equines that may also be reproduced through nuclear transplantation using the horse as the donor oocyte and as the recipient animal carrying the zygote to term.

5.14 PRACTICAL APPLICATION OF CLONING IN THE EQUINE

The technology has broad application to horses. Because of the often high unit value of the horse compared to other agriculture species, the horse may be a species in which cloning has the most immediate application. The application of this technology may be at efficiencies in the horse that are currently unacceptable in other species. Of further significance may be the lower postimplantation loss currently experienced in the horse compared to that of other domestic species. However, additional research and data are necessary before this conclusion can be fully justified.

As of this writing, four healthy, viable, strong, and normally growing (based on all parameters measured) equines have been produced by nuclear transfer. None of these animals has been reported or identified as having any abnormal condition or developmental state; from all appearances, these animals are indistinguishable from animals produced by conventional breeding programs. Obviously, additional animals must be produced to confirm these observations; however, the lack of any detectable neonatal or subsequent growth problems is encouraging and may suggest that the equine is a species that will have a reduced rate of postpartum problems associated with cloning.

The application of cloning technology as a tool for reproduction of select animals may depend on acceptance of the technology by the horse industry, rather than efficiency of the technology in production of offspring. Traditionally, this industry has been slow to incorporate emerging technologies compared to industries representing other species of agriculturally important animals. The widespread use of artificial insemination (AI) in this species did not occur until 20 years after this practice was implemented and commonly used in most other agricultural animals.

Of horse breed associations that allow AI, it has only been in approximately the last 10 years that the stallion producing the ejaculate did not need to be present on the same farm as the mare receiving the insemination. Indeed, only in the recent past have a limited number of breed associations allowed use of "shipped cooled" semen for insemination of mares. Currently, the Thoroughbred Association (racing horses) continues to allow registration only of foals produced from "natural" breeding. Therefore, use of cloning could be slow to evolve as an assisted reproduction tool in the horse industry.

REFERENCES

1. Woods, G.L. et al., A mule cloned from fetal cells by nuclear transfer, *Science*, 301(5636), 1063, 2003.
2. Galli, C. et al., A cloned horse born to its dam twin, *Nature*, 424, 635, 2003.
3. Hinrichs, K., Production of embryos by assisted reproduction in the horse, *Theriogenology*, 49, 13, 1998.
4. Allen, W.R., Kydd, J.H., and Antczak, D.F., Interspecies and extraspecies equine pregnancies, in *Equine Reproduction*, McKinnon, A.O. and Voss, J.L., Eds., Lea & Febiger, Philadelphia, 1993, Chap. 63.
5. Long, C.R. et al., New commercial opportunities for advanced reproductive technologies in horses, wildlife, and companion animals, *Theriogenology*, 59(1), 139, 2003.
6. Cibelli, J.B. et al., Cloned transgenic calves produced from nonquiescent fetal fibroblasts, *Science*, 280(5367), 1256, 1998.
7. White, K.L. et al., Establishment of pregnancy after the transfer of nuclear transfer embryos produced from the fusion of argali (*ovis ammon*) nuclei into domestic sheep (*ovis aries*) enucleated oocytes, *Cloning* 1(1), 47, 1999.
8. Kline, D. and Kline, J.T., Repetitive calcium transients and the role of calcium exocytosis and cell cycle activation in the mouse egg, *Dev. Biol.*, 149, 80, 1992.
9. Campbell, K., Reed, W.A., and White, K.L., The ability of integrins to mediate fertilization, intracellular calcium release, and parthenogenic development in bovine oocytes, *Biol. Reprod.*, 62, 1702, 2000.
10. Vanderwall, D.K. et al., Equine cloning: current status and initial findings on embryonic, fetal and placental development; parturition; and neonatal foal health of cloned mules, *JAVMA*, 225, 1694, 2004.
11. Allen, W.R. and Moor, R.M., The origin of the equine endometrial cups. I. Production of PMSG by fetal trophoblast cells, *J. Reprod. Fert.*, 29, 313, 1972.
12. Allen, W.R., Hamilton, D.W., and Moor, R.M., The origin of the equine endometrial cups: II. Invasion of the endometrium by trophoblast, *Anat. Rec.*, 177, 475, 1973.
13. Hill, J.R. et al., Evidence for placental abnormality as the major cause of mortality in first-trimester somatic cell cloned bovine fetuses, *Biol Reprod.*, 63(6), 1787, 2000.
14. De Sousa, P.A. et al., Evaluation of gestational deficiencies in cloned sheep fetuses and placentae, *Biol Reprod.*, 65(1), 23, 2001.

6 Cloning in the Rabbit: Present Situation and Prospects

*Patrick Chesné, Peter Chrenek,
Mireille Challah–Jacques, Nathalie Daniel,
Laurent Boulanger, Xavier Vignon,
Pierre G. Adenot, and Jean-Paul Renard*

CONTENTS

6.1 INTRODUCTION

Among the species used today in biomedical research or for pharmacology and toxicology studies, the rabbit offers several main advantages over the widely used rodent models. First, its larger size allows several physiological manipulations to be repeatedly performed more easily than in small-sized mice [1]. Second, most if not all toxicology objectives related to reproductive physiology can include longitudinal analysis because semen and embryos can be obtained from live animals (see the review in Foote and Carney [2]), which is not feasible in rodents. Third, a large historical control database exists in the rabbit to distinguish treatment-induced fetal anomalies from naturally occurring ones [3]. Fourth, the biology and physiology of the rabbit have been characterized in detail since this species was first used as a pioneer model at the emergence of mammalian experimental embryology at the turn of the 20th century [4]. Finally, phylogenetic and genetic comparisons of different mammalian genomes have clearly shown that rabbit is closer to human than mouse is [5].

As a consequence, the rabbit is often considered as a pertinent animal model for specific questioning in basic or medical research. For instance, new functional properties of the human class I histocompatibility complex have been discovered from the functional analysis of the rabbit gene, which has exon and intron organization very close to the corresponding human gene [6]. Atherosclerosis and cardiomyopathy are two examples of medical areas that prioritize the rabbit model rather than the rodent model [7]. Because the structure of the CFTR gene (whose mutation in humans results in cystic fibrosis) is very similar in the rabbit [8], this species is considered a better model than the mouse in this research area.

Another example of the potential of rabbit in biomedical research has been recently provided by Chen and coworkers [9], who showed that the cytoplasm of rabbit oooytes is able to support the *in vitro* development through to blastocyst of embryos reconstituted with human nuclei. Although a matter of ethical and scientific debate, this innovative approach opens new possibilities for the production of human stem-like cells from given patients and their use as a screen in a pharmacogenetic optimization of therapeutical treatments.

For several years the rabbit species has also proven to be a very valuable species for producing recombinant proteins from the milk of transgenic females [10–12]. This is because it is possible to obtain enough product from the milk of genetically modified females in a much shorter period of time than with larger species (sheep, cattle). Today, however, the creation of genetically modified rabbit lines is largely hampered by the lack of suitable technologies for controlled modifications of the genome, as is the case in the mouse. In the mouse, targeted germline modifications are routinely obtained from chimeric embryos produced with genetically modified embryonic stem cells. Several years ago, some reports claimed that pluripotent rabbit cell lines can be isolated from embryos to produce chimeric offspring [13]; however, no evidence indicates that this approach can lead to success in this species.

Cloning, i.e., the transfer of the nucleus of a genetically modified somatic cell into a recipient enucleated oocyte, offers an attractive alternative strategy to produce targeted genetic modifications. The principle of this approach was first proven in sheep [14] and more recently in cattle [15]; however, until now it had not been established in rabbit despite the pioneering achievement in this species with nuclei isolated from early cleaving embryos [16–20]. Thus, prospects for the future of genetic modifications in rabbits [12,21] should be viewed with caution.

Recently, however, evidence has been provided that live offspring could also be obtained from rabbit somatic donor cells [22]. Despite this breakthrough, the efficiency of this nuclear transfer technique has remained low with less than 1% of the reconstructed embryos able to develop into normal young [23].

This chapter reviews several specificities of the rabbit with respect to nuclear transfer and explains the technical approaches that have allowed generation of viable and fertile offspring. It also provides an analysis of the limits in reprogramming efficiency in comparison with other species. Finally, some ongoing efforts to overcome difficulties still encountered in this species are reported.

6.2 RECONSTRUCTION OF RABBIT EMBRYOS THROUGH NUCLEAR TRANSFER

Oocytes are obtained after hormonal treatment from superovulated females mated with a vasectomized male to induce ovulation. Ovulated oocytes at the metaphase II stage are recovered 16 h after mating by flushing the oviducts. Then, follicular cells surrounding oocytes are removed after enzymatic treatment and pipetting. Enucleation is performed after staining with a vital DNA dye (Hoechst 33342) by means of an inverted microscope equipped with micromanipulators, under UV light control (see Challah–Jacques et al. [23]). During oocyte manipulations, three points seem to be of great importance:

- Oocytes are very sensitive to the ambient temperature; they must be handled as quickly as possible above 25°C during the removal of *corona* cells, with the minimum of thermal variations to avoid uncontrolled activation.
- Special care should be taken for enucleation in order to avoid oocyte chromatin artifactually participating in embryonic development. In the present case, the activation protocol (also used for SCNT procedure) is able to lead to 89% blastocysts from nonenucleated metaphasic oocytes (see the two last lines in Table 6.1).
- The same care must be applied to remove the polar body during the NT manipulation procedure. Indeed, in a previous experiment from enucleated oocytes fused with polar body and subjected to the activation protocol, 30% developed through to the blastocyst stage.

Blastomeres were first used as a source of donor cells in order to confirm previous successes of several years ago [18,24]. Individual blastomeres (donor cells) were dissociated from morula stage embryos recovered *in vivo* from superovulated does. Each of them was micromanipulated individually and placed with the help of a micropipette beneath the zona of an enucleated oocyte (cytoplast). Direct injection

TABLE 6.1
Effect of Different Treatments on *In Vitro* Development after Parthenogenetic Activation of Rabbit Oocytes

Treatment First Electrical Activation 1-h Incubation in M199 Second Electrical Activation and	No. Oocytes Used	No. Oocytes Cleaved (%)	No. Blastocysts (%)
Cycloheximide 30 min	48	41 (85.4)	19 (39.6)
Cycloheximide 1 h	48	45 (93.7)	24 (50.0)
Cycloheximide 4 h	48	47 (97.9)	25 (52.1)
6-DMAP 30 min	48	43 (89.6)	32 (66.7)
6-DMAP 1 h	48	46 (95.8)	38 (79.2)
6-DMAP 2 h	48	47 (97.9)	35 (72.9)
CHX/6-DMAP 30 min	52	45 (93.7)	37 (71.1)
CHX/6-DMAP 1 h	130	130 (100.0)	116 (89.6)
CHX/6-DMAP 1 h (reconstructed oocytes[a])	36	36 (100.0)	32 (88.9)

[a] MII-karyoplast exchange and removing of polar body by means of micromanipulation to evaluate the NT procedure effect.

of nucleus isolated into the cytoplast can also lead to full-term development as first demonstrated in mouse [25]; until now, however, this technique has not succeeded in rabbit.

Electrofusion of a donor cell and cytoplast was done in a fusion chamber between two electrodes in nonelectrolytic solution (mannitol) with a set of electric pulses generated by means of an electrostimulator. These electrical pulses are not able to activate embryos reconstructed from nonaged oocytes (see Section 6.3.2).

The approach was extended by using cumulus cells surrounding ovulated oocytes collected from another donor female as donor cells. Cumulus cells were first chosen as a source of nuclei because comparative studies between different species suggested that cumulus cell-derived cloned embryos have the best potential to develop *in vivo* [25] (see the section on nuclear reprogramming also). Fibroblastic cells cultured from fetal tissues were also used as donor cells because they can be cultured and seem to support many passages and in vitro genetic modifications.

Reconstructed fused embryos are subjected to an activation protocol that induces a necessary remodeling and reprogramming of the nucleus — the first important steps towards a development state (see the sections on activation and nuclear reprogramming).

After electrofusion (see Chesné et al. [22] and Challah–Jacques et al. [23]), reconstructed embryos are cultured before activation for 1 h in M199 supplemented with 10% FCS at 38.5°C under 5% CO_2 in air. Under these conditions, exposure of fused nuclei to a metaphasic cytoplast leads to nuclear envelope breakdown and chromatin condensation induced by a high level of MPF activity (Figure 6.1A and

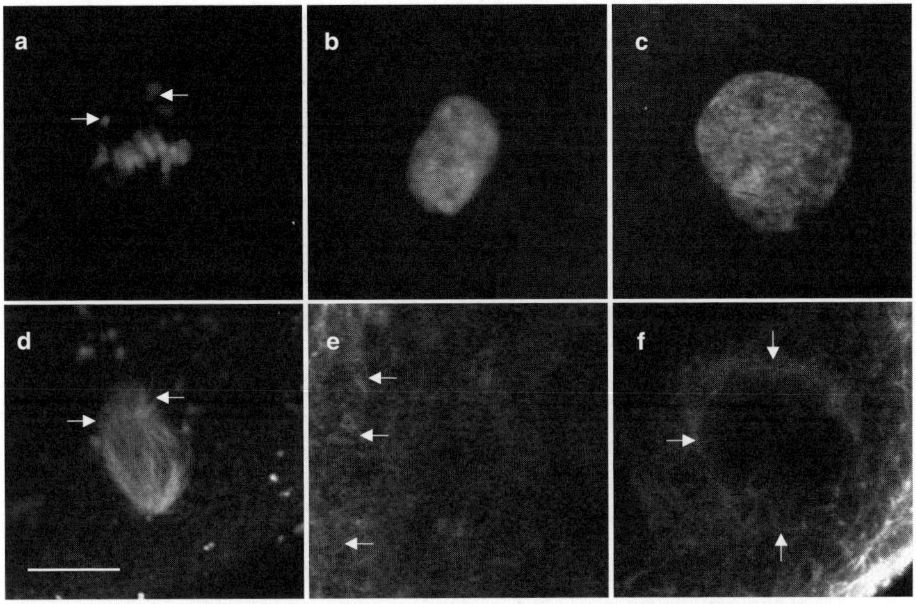

FIGURE 6.1 Confocal microscopy images of DNA (A to C) and microtubules (D to F) in one-cell NT embryos stained with propidium iodide and immunolabelled with an anti-α-tubulin antibody. (A,D) Following fusion, NT embryos were cultured for 1 h to induce PCC before the second set of electrostimulation required for egg activation; PCC led frequently to a misaligned metaphase plate associated with the spindle, and sometimes individual chromosomes (A, arrows) were localized near the spindle poles (D, arrows). (B,E) After the second set of electrostimulation, NT embryos were incubated for 1 h in CHX and 6-DMAP; upon the removal of drugs, 72% of NT embryos (n = 25) already showed a small nucleus and an interphasic microtubular network formed (E, arrows). (C to F) One hour later, all NT embryos were in interphase and 71% (n = 17) exhibited a single and large pronucleus-like nucleus such as that observed in rabbit zygotes (not shown) surrounded by microtubules (F, arrows). Bar = 10 μm.

Figure 6.1D). This situation does not seem to be detrimental for nuclei in G_1 of the cell cycle. It enables chromatin exposure to cytoplasmic components involved in the remodeling and reprogramming of somatic cell nuclei as suggested by Wells et al. [26] and is able to support a high rate of development to the blastocyst stage.

After a second set of electrical pulses, reconstructed embryos are incubated for 1 h in the presence of cycloheximide (CHX) and 6-dimethylaminopurine (6-DMAP). These two drugs are known to permit the passage from a metaphasic to interphasic cytoplasmic environment and remodeling of nucleus (Figure 6.1B and Figure 6.1E; also, see the section on activation).

6.3 OOCYTE ACTIVATION IN THE RABBIT

Activation refers to the process that allows an oocyte arrested at the metaphase of the second meiotic division to enter into interphase. Several protocols have been proposed to try to mimic what happens in the fertilized egg. For that, several stimuli such as electrical pulses, ionophores, and specific molecules like inositol 1,4,5-triphosphate (InsP3) were combined to disrupt the calcium pathway with drugs that mostly show a broad-spectrum effect to inhibit protein synthesis and/or kinase activities.

6.3.1 Activation of Fertilized Embryos

In the fertilized egg, increases in Ca^{2+} induced by the incoming sperm through the phospholipase C/InsP3 pathway are essential for cortical granule exocytosis; meiosis resumption from metaphase II (MII) by stimulating Ca^{2+}/calmodulin-dependent kinase II activity (CDK); and cyclin B1 degradation by the anaphase-promoting complex/cyclosome (APC/C)-dependent process until pronuclei are formed [27] (reviewed by Jones [28]). Consequently, CDK1 activity of the M-phase promoting factor (MPF), which is responsible for chromosome condensation in all metaphasic cells, drops rapidly to a basal level and the fertilized egg extrudes a second polar body containing a haploid set of maternal chromosomes. These processes interact with the MOS/MEK1/MAPK/p90rsk signaling pathway, known as the cytostatic factor (CSF), and inactivation of these kinases leads to MII exit; APC/C-dependent cyclin degradation; loss of the spindle integrity; and pronuclear envelope assembly [29].

Whether the MPF/CSF machinery interferes with sperm chromatin remodeling activities present in the oocyte has not yet been studied. Protamine/histone replacement in sperm chromatin before pronuclear formation depends on removal activities present after germinal vesicle breakdown and chromatin condensation in the maturing oocyte. This replacement requires the maturing oocyte to synthesize nonhistone proteins, which directly or indirectly drive histone transfer activity (reviewed by McLay and Clarke [30]). They are much more efficient during oocyte activation, probably ATP-dependent, and may be regulated through a Ca^{2+}-dependent process.

6.3.2 Activation of Reconstructed Embryos

Physiological equivalences between induced oocyte activation (of the reconstructed embryo) and spontaneous activation (of the fertilized egg) are still poorly determined (see, for instance, Fulka et al. [31]). Parthenotes differed from fertilized eggs in the absence of paternal chromatin and also in several subtle molecular differences, as shown in the mouse [32] where maternal chromatin exhibited acetylated histones H4-K5, before pronuclear formation in parthenogenotes but not in zygotes. Thus, the fact that all activation protocols only partially mimicked what happens in fertilized eggs should always be kept in mind. This has been elegantly demonstrated in the rabbit, in which application of different electrical pulse regimes to the oocyte results in differences in their ability to develop to the blastocyst stage — as well as much further — during postimplantation development [33,34].

Thus, the calcium transients induced by electrostimulation of the oocyte can have long-lasting effects on development, thereby showing the importance of early epigenetic modification on the latter patterning of the embryos. These experimental approaches open new perspectives for a better understanding of basic aspects of cellular and developmental physiology [35]; however, they also show that the extent to which a somatic chromatin can be reformatted into a zygotic chromatin by a controlled electrophysiological process remains a challenging question for cloning.

In the rabbit, freshly ovulated oocytes (14 to 15 h after LH-HCG stimulation), which correspond to "normal" fertilizable oocytes, are poorly activated following Ca^{2+} stimulation except when stimuli are frequently repeated [33,34,36–38]. A short period of "aging" resulting from a 4- to 5-h delay between oocyte recovery and stimulation (*in vitro* aging) or a 2-h delay at oocyte collection (*in vivo* aging) improved activation, but significant development still required repetitive Ca^{2+} stimuli [16,17,19,20,39–42]. A longer delay of *in vitro* (8 to 10 h) or *in vivo* (4 to 5 h) aging is required to bypass this limitation. However, a period of aging also prevents premature chromosome condensation (PCC) after nuclear transfer because oocytes are already in interphase 1 h after Ca^{2+} stimulation. This short timing at the onset of oocyte activation is probably due to rapid inactivation of kinases involved in the MII arrest (see Adenot et al. [32] and references listed there). It also occurs when oocytes aged for a short time are incubated in the presence of 6-DMAP, a broad-spectrum protein kinase inhibitor, following activation stimuli [22].

After a short period of oocyte aging, the transition from MII to interphase in the absence of kinase inhibitor lasts 3 to 4 h [39, 43] — similar to that reported with "normal" oocytes [33]. However, oocytes aged for a short period are known to develop poorly after fertilization; those aged for a long period do not. For the successful development of reconstructed embryos, the following dilemma remains: although oocytes similar to those that support full-term development after fertilization are not suitable for nuclear transfer, aged oocytes can support early development after somatic nuclear transfer but may have already lost some of the properties required for full-term development fertilizable oocytes. Somatic rabbit cloning laboratories are now using freshly ovulated or oocytes aged for a short time that have been activated with a combination of Ca^{2+} stimuli and drugs with the aim to induce a normal timing of early developmental events (cleavage and blastocyst formation) and an efficient rate of oocyte activation following nuclear transfer [22,36,41,44–46].

Similar activation protocols are used today with several mammalian species. However, in contrast to those of other mammalian species, the rabbit zygote has the particularity of entering the S phase very early after activation [47,48]. Therefore, the authors considered that the duration of the MII to interphase transition should be taken into account when establishing an activation protocol for nuclear transfer in this species. Therefore, the protein synthesis inhibitor cycloheximide (CHX) and the kinase inhibitor 6-DMAP were tested to determine whether, alone or combined, they should be used in the activation protocol (Table 6.1, unpublished results).

It was found that a 30-min incubation with CHX, 6-DMAP alone, or with both drugs is already efficient in activating parthenotes and in obtaining 40, 67, and 71% of blastocyst rates, respectively. A 1-h treatment with both drugs gives 90% of blastocyst stage parthenotes. This incubation period with drugs was used for successful somatic rabbit cloning [22]. In NT embryos, cytoplasmic exposure in metaphasic environment of fused nuclei leads to the breakdown of nuclear envelope and chromatin condensation (Figure 6.1A and Figure 6.1D). Following incubation for 1 h with drugs, a nucleus is formed and an interphasic microtubule network assembled (Figure 6.1B and Figure 6.1E). One hour after the removal of drugs, a single, large, pronucleus-like nucleus surrounded by interphasic microtubules is observed (Figure 6.1C and Figure 6.1F), thus indicating that the somatic nucleus has been reprogrammed and has already entered in the one-cell embryonic cycle.

6.4 CELLULAR AND MOLECULAR EVIDENCE OF NUCLEAR REPROGRAMMING IN THE RABBIT

The first evaluation of nuclear remodeling in the rabbit appeared in studies in which it was shown that the time required to progress from the zygotic to the blastocyst stages was similar between fertilized and nuclear transfer embryos. In the rabbit, this period lasts approximately 120 h when using eight-cell stage blastomeres as a source of nuclei [16]. These nuclei were thus apparently correctly reprogrammed because they reverted to the same morphological and temporal developmental pattern as the zygote.

Another indication of nuclear reprogramming is RNA synthetic activity. The fusion of both transcriptionally silent gametes is followed by a period of absence of gene expression and then a period of permissive expression. The transition from maternal to zygotic control was described at the 8- to 16-cell stage rabbit embryo [49]. At this stage, all blastomeres are transcriptionally active and synthesize mRNA and rRNA [50]. After transplantation of 32-cell stage morula nuclei in enucleated oocytes by electrofusion, the synthesis of nuclear RNAs stops independently of the cell cycle phase of the donor nucleus within 2 to 12 h after fusion and is weakly reactivated between the two-cell and four-cell stage. The transcription activity then rapidly increases through the 16-cell stage, reaching the level typical of 32-cell stage nuclei in normal rabbit development [50]. Thus, the cytoplasm of the enucleated oocyte is able to block transcriptional activity of the donor nucleus and to trigger a new RNA synthesis in nuclear transfer embryos.

These basic observations, however, do not prove that a true reprogramming has taken place because only a small proportion of cloned embryos of rabbit or other species develops to term. The chromatin remodeling and reprogramming processes in embryonic development can also be assessed through the evaluation of genomic methylation, which largely participates in the epigenetic regulation in cells. Recent observations point to the importance of DNA methylation, which is one of the epigenetic parameters involved in genome regulation.

6.4.1 Epigenetic Reprogramming of Donor DNA after Nuclear Transfer

A complete reprogramming of a somatic nucleus by recipient cytoplasm would result in an embryo with the same profile of gene transcription as that seen *in vivo* or even after *in vitro* fertilization. The expression profile of candidate genes provides a first estimation of the extent to which the donor genome is correctly reprogrammed or not. This can be monitored in preimplantation embryos by the timing of the onset of transcription of genes not expressed in the donor somatic cells but required for embryonic development, or by the switching off genes transcribed in the donor somatic cell but not normally transcribed in early embryonic development. With this latter approach, it was shown that the mRNA profiles of bovine blastocysts obtained by nuclear transfer of fetal fibroblasts were highly conserved and similar to those of blastocysts developed *in vivo* or *in vitro* [51].

More extensive studies were performed with the former approach, in which the expression of some developmentally important genes has been investigated in nuclear transfer as compared to their *in vitro* or *in vivo* counterparts. Altered levels of transcripts were found for several growth factors and cytokine genes in nuclear transfer-generated embryos [52], and the relative abundance of mRNA in nuclear transfer embryos was shown to be altered by the activation protocol, as well as the types of donor cells [53].

Aberrant gene expression patterns after nuclear transfer may be involved in the loss of embryos and fetuses during pregnancy or at perinatal stages [54]. It therefore appears that even though somatic (or blastomere) genes are switched off upon nuclear transfer, the genome of the nuclear transfer embryo is not exactly regulated. It suggests more generally that the kinetics of nuclear remodeling events during the early embryonic stages is a key factor for the onset of gene regulation.

In this context, the rabbit species provides an attractive model for studies on nuclear reprogramming and gene regulation in preimplantation embryos. First, the maternal to embryonic transition occurs several cell cycles after fertilization, thus providing a large window of time in which processes related to the progressive gene reactivation can be investigated. Second, these events can be quickly correlated with full-term development, which is much shorter in rabbit than in other large animals. Third, some genes in rabbit embryos that are transiently transcribed at the 8/16 cell stage have been identified; this makes them good candidates to study the potential alterations of developmentally regulated genes in relation to nuclear transfer technology [55].

Most genes in differentiated cells are regulated through epigenetic modifications. The epigenetic modification of DNA in the promoter region of genes is correlated to gene silencing and is associated with changes in the acetylation of nucleosomal histone, which may in turn modulate gene expression (review Wade and Kikyo [56]). Epigenetic information provides instructions on how, where, and when the genetic information should be used. It does not involve any change to the DNA code and can persist through one or more generations. A molecular hypothesis put forward to explain the common mechanisms underlying growth and other abnormalities as a consequence of cloning by nuclear transfer is that this procedure or the culture of preimplantation embryos and cells can lead to epigenetic modifications in genes, which would affect their expression at later stages of development [57–59]. One of the most studied epigenetic regulations of gene expression in mammals is DNA methylation, which could be the cause of deregulation of development when it is modified by external factors [57]. This implication of methylation in abnormal development finds support in the fact that imprinted genes can suffer alteration of expression upon prolonged periods of culture in mouse ES-cells or embryos [57,59]. It is generally assumed that reprogramming is associated with an active demethylation of paternal DNA and a passive demethylation in the embryos as the DNA replicates up to morula stages [60].

Based on methylcytosine staining, aberrant or incomplete reprogramming was evidenced in bovine cloned embryos [61,62]. This alteration was associated with aberrant gene expression and developmental abnormalities during the postimplantation stages of these nuclear transfer embryos [61]. In contrast, neither active demethylation at the one-cell stage nor further passive demethylation was observed in the rabbit embryos [63,64]. This suggests that genome-wide demethylation is not a prerequisite for normal development and that more subtle changes are linked to the reprogramming process.

In rabbit, the gene reactivation preceding the maternal to embryonic transition corresponds to a developmental period in which chromatin is progressively remodeled, resulting in the progressive establishment of a finely tuned transcriptional regulation [65]. The rabbit species would be an interesting model to study if subtle DNA reorganization undetectable by immunostaining is required for normal development and to determine which region of the genome must be reset during this period in nuclear transfer embryos to better explain aberrant reprogramming.

6.4.2 Importance of Donor Cells

6.4.2.1 Cell Cycle Stage

Numerous studies in sheep, cattle, mouse, and goat claimed that the use of G_0 cells for embryo reconstruction is beneficial to the development of somatic nuclear transfer embryos [25,26,66–68]. Conversely, in other studies, no difference was found in the rate of blastocysts obtained with serum-starved (G_0) cells or non-serum-starved cells [69–71] with adult cells. In rabbit, serum starvation of donor cultured fibroblasts did not improve the rate of development to the blastocyst stage and no viable offspring was obtained whatever the culture conditions of the donor cells [36].

It is still debatable which cell cycle stage, G_0 or G_1, results in the best cloning efficiency. The G_0 stage per se seems to have no positive effect on the developmental potential of nuclear transfer embryos in cattle, pigs, and even mice [72–74]. Moreover, cells in G_1 at the moment of nuclear transfer in bovine supported higher developmental rates *in vivo* [75]. In rabbit, development to blastocysts is greatly affected by the stage of the cycle of the donor nucleus when using embryonic cells. Blastomere cycle synchronization appeared to improve cloning efficiency; a beneficial effect of using early stage donor blastomeres was evidenced with donor nuclei in the G_1 phase [76]. Notably, donor cells in S-phase led to chromosome and spindle disorganization in most cases of PCC in nuclear transfer embryos and to a lower development to the blastocyst stage [40,77].

The fusion of 17-19hpc aged oocytes with embryonic cells or G_0/G_1 somatic cells for embryo reconstruction proved to be the best combination for an optimal rate of *in vitro* development [22,40,78]. However, the *in vivo* development remained relatively poorly efficient except for embryonic cloning. Thus suggesting that cell cycle synchronization only affects the blastocyst rate and that the full reprogramming depends more on the type of cell used as donor.

6.4.2.2 Donor Cell Type and Further Development of Nuclear-Transferred Embryos

Various types of differentiated cells have been used as a source of nuclei for cloning domestic and laboratory animals. The different cells available do not exhibit the same potential for further development of the reconstructed embryos. Cumulus and granulosa cells have been extensively used for nuclear transfer in mice, cattle, and pigs. Comparative studies suggest that cumulus cell-derived cloned embryos have a better potential to develop *in vitro* and *in vivo* in these species [25,79–81].

In rabbit also, *in vitro* development to the blastocyst stage was affected by donor cell type. Rate of development with adult fibroblasts ranged from 6 to 18% [36] to 30% [44] and from 6 to 40% with fetal fibroblasts [82]. When cumulus cells were used, developmental rates were consistently higher, with cultured cumulus cells (23% to the blastocyst stage [41]) and with fresh cumulus cells (47% [22]). In a comparative study, Cervera and Garcia–Ximenez [78]. also evidenced a better development to the blastocyst stage when using cumulus cells (7%) as compared to fetal fibroblasts (1%). Although great variability in cloning efficiency can be observed among species that have been successfully nuclear transferred, almost all donor cell types tested resulted in live offspring [80,83].

Conversely, only fresh cumulus cells allowed full-term development of nuclear transfer embryos in rabbit [22]. This specificity of rabbit with regard to full-term survival points out the need for a better understanding of the regulation of implantation and fetal development processes in this species. The reason why cumulus cells give the highest cloning efficiency and the lowest number of abnormalities in cloned animals is not elucidated. It may be related to a more efficient reprogramming process of the DNA of cumulus cells, as suggested by studies conducted in mice, which showed that cumulus cell-derived cloned embryos do not have widespread dysregulation of imprinting [84]. If a less differentiated state of cumulus cells is responsible for this better ability to nuclear reprogram, then the use of pluripotent cells should lead to an improved development of nuclear transfer embryos. This is the case with embryonic cells isolated from morulae (see Section 6.2). However, neither rabbit ES-like cells derived from ICM of 3- to 4-day-old rabbit embryos nor primordial germ cells (PGC) isolated from gonads of 18- to 22-day-old fetuses could yield implanted fetuses [85,86].

The loss of developmental potential observed when using cultured somatic cells as compared to fresh cells might be associated with culture-induced change in the donor cell lines. Nuclei from cultured cells may lose their ability to support full and normal embryonic development with time because of possible accumulation of genetic or epigenetic alterations. Such genomic instability has been demonstrated in cultured ES-cells in which changes in imprinted gene expression could be observed in cells coming from the same subclone [87]. This may contribute to the drop in implantation observed when mice nuclear transfer embryos were derived from ES-cells with more than 19 passages [88]. It has also been evidenced in mouse ES-cells that culture conditions and, especially serum starvation or culture at high cell density, could induce alterations to methylation and expression patterns of the imprinted genes Igf2 and H19 [89]. This could in turn affect the embryonic and fetal development of nuclear transfer embryos derived from *in vitro* cultured donor cells.

6.5 *IN VITRO* DEVELOPMENT OF CLONED RABBIT EMBRYOS

Media commonly used by different authors for *in vitro* culture of rabbit embryos are M199, Ham's F10, KSOM, EBSS, M2 supplemented with serum or bovine albumin. In the authors' experiments, short time incubations of oocytes and reconstructed embryos are done in M199 supplemented with 10% FCS at 38.5°C under 5% CO_2 in air. Nuclear-transferred embryos are then cultured overnight in 30 μl microdrop of B2 medium (CCD Laboratories, Paris, France) supplemented with 2.5% FCS under mineral oil at 38.5°C under 5% CO_2 in air, until the four-cell stage before transplantation into recipient foster mothers or until blastocyst stage for evaluation. This medium currently leads to 92.8% development until the blastocyst stage after culture of control one-cell stage zygotes during 3 to 4 days.

Under these conditions, a large proportion of the embryos reconstructed from cumulus but also from fibroblast cells cleave (more than 80%), and nearly half of them routinely developed into expanding blastocyst at day 4. An analysis of total cell number at two different periods of blastocyst development (day 3 and day 4) showed, however, that despite their apparently normal morphology, the development of reconstructed blastocysts is somewhat delayed in comparison to fertilized embryos or parthenotes (Table 6.2).

At day 3, the mean cell number of embryos reconstructed with freshly recovered (cumulus) or cultured (fibroblast) donor cells is similar; however, in both cases, it is already behind that of fertilized embryos. At day 4, the development of blastocysts obtained from embryos reconstructed with cultured somatic cells (fibroblasts) lags behind those reconstructed with fresh cumulus cells; in both cases, however, embryos are still developing because the total number of cells increases between day 3 and day 4.

TABLE 6.2
Mean Cell Number of Rabbit Blastocysts Obtained after *In Vitro* Culture during Days 3 and 4

Blastocysts Obtained from	Blastocyst No. (%)[a]	Day 3 (mean ± s.e.)	Day 4 (mean ± s.e.)
Zygotes cultured *in vitro* from 1-cell stage	52/56 (92.8)	90 ± 19 (n = 17)	193 ± 52 (n = 27)
Parthenotes (activated oocytes)	116/130 (89.6)	73 ± 16 (n = 19)	133 ± 41 (n = 19)
NT embryos reconstructed from cumulus cell nuclei	215/474 (45.4)	52 ± 18 (n = 16)	94 ± 19 (n = 15)
NT embryos reconstructed from fibroblast nuclei	89/203 (43.8)	54 ± 15 (n = 11)	70 ± 25 (n = 38)

[a] Number of blastocysts/number of one-cell stage embryos.

The extent to which this difference between the two types of donor cells depends on differences in the reprogramming behavior of their chromatin or on differences in the requirement of the reconstructed embryos during culture becomes an important question in nuclear transfer research. This is because of the important observation that, in the mouse, the somatic program of gene expression of the donor nucleus remains for a while after nuclear transfer [90]. Thus, the metabolic activities of the reconstructed embryo can differ somewhat, at least at the beginning, from normal fertilized ones. This may have important consequences for the preimplantation development of rabbit embryos (*in vitro* and also *in vivo*) because of the rapid growth of the blastocyst before implantation in that species.

It has been shown that two types of rabbit genes were turned on at the onset of zygotic activity at the 8- to 16-cell stage: one corresponds to genes expressed transiently and the other to genes expressed in the long term [55,91] and includes several mitochondrial genes with rapidly increasing levels of gene expression (authors' unpublished observations).The role of microenvironment on the developmental kinetics of reconstructed rabbit embryos and its consequences on postimplantation development are now under investigation in the authors' laboratory using appropriate genomic tools (Duranthon et al., unpublished data).

6.6 *IN VIVO* DEVELOPMENT OF RABBIT EMBRYOS RECONSTRUCTED FROM SOMATIC CELLS

In rabbit species, the degree of synchrony between the physiological status of the recipient and the developmental stage of the embryo at the time of transfer is important because of the narrow window for implantation in this species. Highest rates of full-term development are obtained with normal fertilized embryos under synchronous conditions — that is, when embryos are placed in the tracts of the recipients at the corresponding corpus luteal stage [92]. Transfer of manipulated (frozen–thawed) rabbit embryos into recipients ovulating after donors is beneficial for the survival rate [93].

In rabbit, implantation is completed at day 7.5 after mating and the presence of glycoprotein layers around the embryo (the mucin coat deposited during the oviducta transit) is an important factor for normal implantation. Mechanical removal of this mucin coat or its absence in embryos cultured *in vitro* from the one-cell stage and transplanted at the blastocyst stage directly in the uterus affects the postimplantation developmental potential of the embryos [94,95]. Thus, transfer of early manipulated embryos must be done as soon as possible, after a short time culture.

Development is poor when NT embryos or parthenogenetically activated oocytes are transplanted through the *infundibulum* at the one-cell stage into recipient females mated with vasectomized males at the same time as donors of oocytes (synchronous recipients). Taking into account the delayed development of NT embryos requires the transplantation into foster mothers during (or even before) ovulation, which in the rabbit occurs approximately 11 h after mating [96] (C. Thibault, personal communication). Surgical transfer of eggs shortly after mating is compatible with term development [97]; however, in the authors' experience, embryo transfer performed at or around the ovulation period dramatically affects the physiology of the recipient. To overcome this barrier, NT embryos were cultured overnight up to the four-cell stage before they were transferred into recipients mated 16 h (or 22 h) after oocyte donors (asynchronous recipient, see Schema 6.1).

It was demonstrated in these conditions that pregnancy rates obtained after transfer of nuclear transfer embryos or parthenotes into asynchronous recipients (–16 h) were increased compared with synchronous transfer, thus providing evidence that the delayed development of nuclear transfer embryos was (at least partially) compensated for by transferring them into an asynchronous environment (Table 6.3). With synchronous transfers, the rate of implantation at day 8 (as assessed after slaughtering recipients) was very low and no morphologically normal blastocyst could be recovered.

These physiological conditions for the transfer of NT embryos into recipients were decisive for the postimplantation developmental potential of rabbit embryos reconstructed with nuclei of two different donor cell types:

- Cumulus cells, which can be cultured only for a few days [41] and had already proven to be a suitable source of nuclei for full-term reprogramming
- Fibroblastic cells, which can be easily grown *in vitro* for periods long enough to be compatible with targeted genetic modifications as shown in sheep or, more recently, in cattle [14,15]

In all cases, pregnancies at midgestation were evaluated by abdominal palpation or echography at day 14.

From 32 asynchronous recipients (–22 h) that were transplanted with four-cell stage embryos reconstructed with cumulus cells (n = 458, 13 embryos/recipient), 11 were diagnosed pregnant after abdominal palpation at day 14 (34.4%) and 5 (15.6%) conducted their pregnancies to term, resulting in the birth of seven live kits (see Table 6.4). In terms of kits born from total embryos transplanted (1.5%, 7 out of 458), efficiency was in the range of those obtained in other domestic mammal species after NT from differentiated cells. The seven living kits were born from 146 embryos transplanted (4.8%) in five pregnant females.

Opening of the 6 (out of 11) recipients that aborted in the second half of pregnancy following bleeding or delivery behavior (pulling out hair) revealed the presence of five dead fetuses and 13 cotyledons and resorbing tissues. These observations indicate that failure in one of the fetuses may lead to complete abortion of the whole litter. In one case, however, the presence of two additional small cotyledons was observed after cesarean section at day 31 and the birth of two live twin kits. Moreover, in this last case, one of the kits was very small (30 g) compared to the other (60 g) and died 2 days later.

Rates of success were even lower when cultured fibroblasts were used as donor cells. When primary culture grown up to 50 days *in vitro* (about eight passages) is used as a source of nuclei, one live kit was obtained after cesarean section (performed at day 32 of pregnancy) (Challah–Jacques et al., to be published). In this experiment (see Table 6.4), nearly 20% of the recipients (19.0%, n = 8) had been diagnosed pregnant at day 14. The live kit was of a normal size at birth (7.5 cm crown–rump)

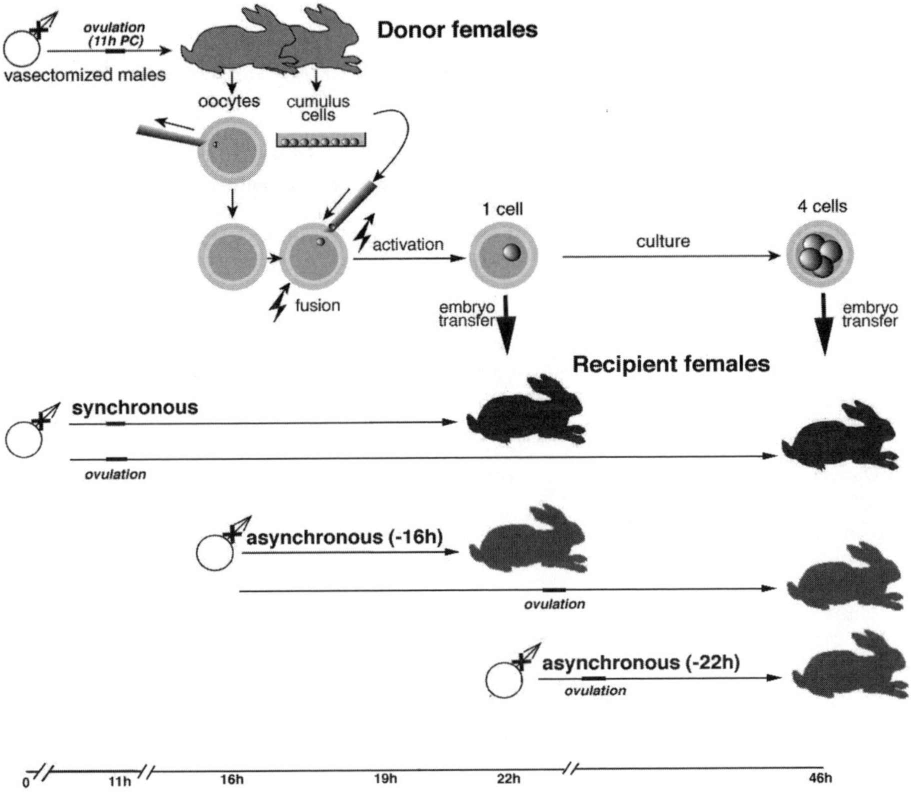

SCHEMA 6.1 Experimental design to assess the effect of synchronization between nuclear transfer embryos and recipient female in rabbit. Oocytes and cumulus cells were recovered from superovulated donor females 16 h postcoïtum (PC) to vasectomized males. Nuclear transfer (NT) procedure (oocyte enucleation, insertion of cumulus cell and electrofusion) was done between 16 and 19 h after mating of donor females. One hour after fusion, activation was accomplished by subjecting NT embryos to a second set of electrical stimulations followed by incubation in the presence of cycloheximide and 6-dimethylaminopurine. In the first experiment, recipient females were mated with vasectomized males at the same time as donor females (synchronous). Nuclear-transferred embryos were transplanted at the one-cell or four-cell stage. In the second experiment, the mating of recipient females to vasectomized males was delayed for 16 h (asynchronous –16 h). Activated oocytes were also transferred to synchronous or asynchronous (–16 h) recipients at the one-cell or four-cell stage. Implantation rate was assessed after slaughtering of recipients at day 8 (see Table 6.3). Mating of recipient females to vasectomized males was delayed for 22 h (asynchronous –22 h). In this set of experiments, four-cell NT embryos were transferred to recipient females and *in vivo* development until term was compared with one-cell stage NT embryos transferred into synchronous recipient.

with a beating heart, but its forebrain was dramatically reduced in size (microcephaly) and it died soon after delivery. Other developmental anomalies were also found on several aborting recipients, in the second half of pregnancy, that were opened

TABLE 6.3
Implantation after Transfer into Synchronous or Asynchronous (–16 h) Rabbit Recipients at Day 8

Type of Embryos Transplanted	Type of Recipients	Pregnant Recipients at Day 8/Total Transferred	No. Implantation Sites/No. Embryos Transferred into Pregnant Recipients (%)	No. Implanted Blastocysts/No. Recovered
Fertilized[a] (one cell)	Synchronous	5/6 (83%)	30/80 (37.5)[a]	9/9
Parthenotes (one cell)	Synchronous	8/20 (40%)	17/78 (21.8)[a]	0/1
NT[b] (one cell)	Synchronous	5/16 (31%)	7/91 (7.7)[b]	0
Parthenotes (four cells)	Asynchronous	5/9 (55%)	15/44 (34.1)[a]	3/3
NT[b] (four cells)	Asynchronous	6/13 (46%)	12/59 (20.3)[a]	1/7

[a] One-cell stage embryos collected from superovulated and fertilized females.
[b] NT embryos from cumulus cells.
Note: Values with different letters were significantly different ($P < 0.05$).

TABLE 6.4
***In Vivo* Development of Rabbit Somatic Nuclear Transferred Embryos: Effect of Donor Cell Culture**

Source of Nuclei	Fibroblasts		Cumulus Cells
Time in culture: no. passages	$33 < P < 40$	$P \leq 8$	None
No. embryos transferred [no. replicates]	323 [24]	765 [52]	458 [36]
No. of recipients: total transferred	23	42	32
Pregnant [% from transferred]			
at day 10		28 [66.7][a]	
at day 14	0	8 [19.0]	11 [34.4]
at day 18		3 [7.1][b]	10 [31.2]
at day 22		2 [4.8][c]	8 [25.0]
at day 28		1 [2.4]	6 [18.7]
Delivering [% from transferred]		1 [2.4]	5 [15.6]
No. kits born [% from embryos transferred]		1 [0.13]	7 [1.5]

[a] Diagnosed after recipient slaughtering.
[b] Four fetuses.
[c] Three fetuses.

between day 22 and day 25. It is thus likely that pregnancies were compromised by a lack of synchrony between the recipient mother and the embryo as well as by misregulations in embryonic development.

No pregnancies could be established after transfer of embryos reconstructed with nuclei isolated from established cell lines. These cells could be kept with a low rate of aneuploidy during culture, and their ability to grow them routinely during about 120 days (up to 40 passages) made them an attractive candidate for targeted genetic modifications. However, the disappointment that resulted from the lack of any positive diagnosis after palpation of recipients at midgestation was a convincing demonstration that much remains to be learned before nuclear transfer can be used for producing rabbit models of human diseases through targeted transgenesis.

6.7 RABBIT CHIMERIC OFFSPRING PRODUCED FROM SCNT EMBRYOS: AN ALTERNATIVE WAY TO REPROGRAM RABBIT FIBROBLASTS UP TO TERM?

The preceding data demonstrate that full-term reprogramming of nuclei from cultured somatic cells is an achievable goal in the rabbit. Meanwhile, they also show that it will be a long time before the first successes achieved in rabbit a long time ago with embryonic nuclei [16] or more recently with somatic nuclei [22] can be turned into applications. These applications deal mainly with the production of genetically modified founder bucks from which new lines of rabbits could be derived and used as animal models in biomedicine, pharmacology, and toxicology. This implies obtaining live offspring as well as ending up with germline transmission of the modified genes after mating.

A recent study in the mouse shows that a simple procedure of clone–clone embryo aggregation at an early stage can improve cloning outcomes [98]. This procedure increases the total cell number of the embryo and can even compensate for deficient expression of genes implicated in peri-implantation development, such as *oct-4*; consequently, the rates of postimplantation and full-term development can be markedly increased. This approach exploits the long-lasting observation that, when precompacted mammalian embryos are brought close to each other, the blastomeres from each embryo will reorganize together to produce a single chimeric embryo [99].

This ability to form blastocysts after aggregation of embryonic cells can even cross species boundaries to produce individual offspring that not only have more than two parents but also have cells from more than two species [100]. It has also been shown, using normal and parthenogenetic blastomeres, that normal embryos are able to stimulate the development of parthenogenetic blastomeres up to term in reconstructed bovine chimeric embryos [101]. The degree of contribution of blastomeres to different lineages in chimeras can thus be used as an assay of the developmental competence of manipulated cells. It can be affected by the genotype of aggregated cells.

The rabbit embryos reconstructed with embryonic cells can result in chimeric animals [102,103]. In this species, a high proportion of embryos reconstructed with different somatic cell types can develop into blastocyst *in vitro* [22,36,44,45,104–107]. Thus, the potential of embryo aggregation was explored not only to produce chimeric offpsring but also, more importantly, to examine whether germline transmission could be obtained from reprogrammed cells.

For this, two different lines of rabbits with distinct phenotypes were used: the New Zealand breed (NZW), which is a white-coated animal, and the Fauve-de-Bourgogne (FdB) with a brown coat. The abilities of 8- to 16-cell stage blastomeres of the FdB breed to contribute to the development of fertilized NZW embryos were compared. Those cloned blastomeres were obtained from cumulus or fibroblast donor nuclei. They can easily be obtained after nuclear transfer, therefore allowing the reconstruction of enough embryos to analyze their contribution in chimeric live offspring, including germline transmission.

6.7.1 Chimera Production (see Schema 6.2)

In this work chimeric young born rabbits were identified by coat color (skin and hair) pattern pigmentation. When two cloned blastomeres (FdB breed) were introduced into host embryos produced *in vivo* from the NZW breed, one would expect

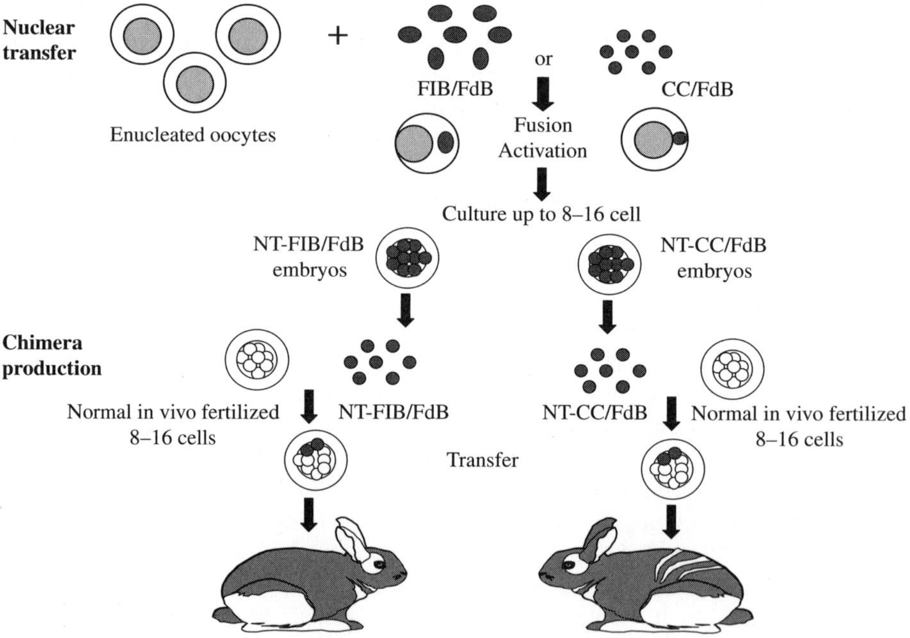

SCHEMA 6.2 Cumulus (NT-CC/FdB) or fetal fibroblasts (NT-FIB/FdB) were fused with enucleated NZW oocytes. The resulting cloned embryos were dissociated into individual blastomeres at the 8- to 16-cell stage; two of these blastomeres were introduced into host NZW embryos produced *in vivo* at the 8- to 16-cell stage (approximately 46 h postfusion) using a pipette with an inner diameter of about 30 μm (Schema 6.2). The chimeric embryos were transferred into oviducts of synchronized pseudopregnant recipients of NZW breed. The number of transferred chimeric embryos per oviduct varied between four and eight, depending on the availability of embryos on a given day.

TABLE 6.5
Chimeric Rabbits Developed from Reconstructed Embryos[a]

No. Replicates	No. Host Embryos[b]	Type of Blastomeres Added to Host Embryos[c]	No. Embryos Transferred/ Recipients (No. Pregnant Recipients)	No. Newborn Rabbits[d]	No. Somatic Chimeric Rabbits (%)[e]	No. Germline Transmission/ Newborn in F1
3	58/N	2/ NT-CC/FdB	58/6 (5)	18/58 (31)	3/18 (17)	0/35
2	65/N	2/ NT-FIB/FdB	65/4 (3)	17/65 (26)	0/17	0/46

[a] Cloned and *in vivo*-produced blastomeres.
[b] N: normal *in vivo* fertilized embryo of the New Zealand breed at 8- to 16-cell stage (host embryo).
[c] NT: blastomeres obtained from 8- to 16-cell stage embryos reconstructed with nuclei of cumulus cells or fibroblastic cells.
[d] Percentage from transferred embryos.
[e] Percentage from kits obtained.

to see overt coat color chimera formation if the cloned cells were capable of participating in normal development. Table 6.5 shows the results.

In the first series of experiments (three replicates), 58 chimeric embryos (2 NT-CC/FdB-derived blastomeres into host 8- to 16-cell stage NZW embryos) were transferred into the oviducts of six recipients, resulting in the birth of 18 living young (Table 6.5). Out of these, three (17%) exhibited a mixed-color coat, with large white bands alternating with colored ones. The degree of chimerism, as judged by the distribution of pigmentation from the cloned FdB cells, was high (see Photo 6.1). Two of these chimeras were females and one was male, but none of the

PHOTO 6.1

35 offspring obtained after the mating of these chimeric animals with pure-breed white NZB had a colored coat.

In the second series of experiments, 65 chimeric embryos obtained from fibroblast-derived blastomeres (2 NT-FIB/FdB-derived blastomeres injected into host NZW embryos) were transferred into the oviducts of four recipients, and 17 live young were obtained and grown to adulthood. No coat pigmentation typical of the FdB breed in the skin or in the hair was detected and no evidence of chimerism could be obtained from the 46 offspring obtained from 17 such animals. These results indicated that the NT-FIB/FdB-derived blastomeres were, in fact, poorly competent for *in vivo* development even when aggregated with normal embryonic cells. Occasionally, however, a chimeric offspring can be obtained (R. Skzryszowska and Z. Smorag, personal communication).

The results of chimeric experiments confirmed that nuclei of cumulus cells supported development up to term. The survival and contribution of cloned blastomeres in chimera offspring are probably influenced to a considerable extent by environmental conditions mediated by the cells from embryos produced *in vivo*. There is evidence for metabolic cooperation between different cell types through permeable cell junctions that enable metabolic deficient cells to function normally [108]. Developing intercellular junctions between blastomeres of different origins might play a role in communication and subsequently enhanced development of the chimeric embryos [109].

Similar interaction between cloned and *in vivo*-produced blastomeres can support cloned cells in the reconstructed embryo. The failure of the NT-FIB/FdB blastomeres to participate in chimera formation reflects their poor ability to reprogram beyond the blastocyst stage [36]. However, this does not prevent further development into fetuses as evidenced from the authors' work (see Table 6.4) and that by others [105,107,110]. The authors' data point to the crucial importance of culture conditions of donor cells because extending their number of passages compromised the postimplantation development of reconstructed embryos. This result confirms the data published in a previous report in which chimeric transgenic fetuses could be obtained using cultured somatic cell nuclear transfer, but only up to about midgestation (15 dpc) [111].

Several key issues should be considered in producing developmentally competent cloned rabbit embryos/blastomeres from fibroblast cells as nuclei donors. First, a direct injection method [25,112], in which the volume of donor cytoplasm incorporated into enucleated oocytes is smaller, might improve the outcome. Second, a double nuclear transfer method [79] in which a karyoplast (containing a reprogrammed pronucleus) is transferred to an enucleated zygote might also improve the developmental competence of rabbit-cloned blastomeres. Third, use of pluripotent cloned embryonic stem cells for chimera production might possibly result in development into fetus if reprogramming of cloned cells is sufficient.

6.8 CONCLUSION

The rabbit has been used for many years in pharmacology and toxicology studies in numerous laboratories. This species is receiving renewed attention today because of its greater proximity to humans than the mouse and the requirement of more than one animal model to validate any new therapeutical procedure or pharmaceutical treatment. The rabbit embryo is also a pertinent model to study basic aspects of early embryonic development; a long tradition of academic research with this species exists (see, for instance, Pincus and Werthessen [113]).

The potential of rabbit in biomedical research has been recently extended by Chen and coworkers [9], who showed that the cytoplasm of rabbit oooytes is able to reprogram human somatic nuclei into pluripotent embryonic-like (ES-like) human stem cells. Although it is a matter of ethical and scientific debate, this innovative approach opens new possibilities for the production of embryonic cells from given patients and their use as a screen in pharmacogenetic studies.

In comparisons with other mammalian species, however, the rabbit appears to be less amenable to reprogramming through nuclear transfer. This impedes the use of this technology to generate genetically modified animals — an awaited perspective in biomedical research. The authors demonstrated a few years ago that rabbit somatic nuclear transfer is an achievable goal and have provided here experimental evidence that improvement in the overall efficiency of this technique remains a reasonable research objective. Together with the rapid enrichment of knowledge on the genetic makeup of the rabbit genome [114], this has already caused consideration of the rabbit as complementary to the mouse in several areas of modern biology and medicine.

REFERENCES

1. Manning, P.J., Ringer, D.H. and Newcomer, C.E., Eds., *The Biology of the Laboratory Rabbit, Academic Press*, San Diego, 1994.
2. Foote, R.H. and Carney, E.W., The rabbit as a model for reproductive and developmental toxicity studies, *Reprod. Toxicol.*, 14(6), 477, 2000.
3. Feussner, E.L. et al., A decade of rabbit fertility data: study of historical control animals, *Teratology*, 46, 349, 1992.
4. Brachet, A., Developpement *in vitro* de blastodermes et de jeunes embryons de mammifères, *C.R. Hebd. Séances Acad. Sci. Paris*, 155, 1191, 1912.
5. Graur, D., Duret, L. and Gouy, M., Phylogenetic position of the order *Lagomorpha* (rabbits, hares and allies), *Nature*, 379, 333, 1996.
6. Marche, P.N. et al., Structure of a functional rabbit class I MHC gene: similarity to human class I genes, *Immunogenetics*, 21, 71, 1985.
7. Fan, J., Challah, M. and Watanabe, T., Transgenic rabbit models for biomedical research: current status, basic methods and future perspectives, *Pathol. Int.*, 49, 583, 1999.
8. Vuillaumier, S. et al., Phylogenetic analysis of cystic fibrosis transmembrane conductance regulator gene in mammalian species argues for the development of a rabbit model for cystic fibrosis, *Mol. Biol. Evol.*, 14, 372, 1997.
9. Chen, Y. et al., Embryonic stem cells generated by nuclear transfer of human somatic nuclei into rabbit oocytes, *Cell Res.*, 13(4), 251, 2003.
10. Stinnackre, M.G. et al., The preparation of recombinant protein from mouse and rabbit milk for biomedical and pharmaceutical studies, in *Transgenic Animals, Generation and Use*, L.M. Houdebine, Ed., Harwood Academic Publishers 1997, 461.

11. Houdebine, L.M., Transgenic animal bioreactors, *Transgenic Res.*, 9, 305, 2000.
12. Bösze, Zs. et al., The transgenic rabbit as model for human diseases and as a source of biologically active recombinant proteins, *Transgenic Res.*, 12, 541, 2003.
13. Schoonjans, L. et al., Pluripotential rabbit embryonic stem (ES) cells are capable of forming overt coat color chimeras following injection into blastocysts, *Mol. Reprod. Dev.*, 45, 439, 1996.
14. McCreath, K.J. et al., Production of gene-targeted sheep by nuclear transfer from cultured somatic cells, *Nature*, 405, 1066, 2000.
15. Kuroiwa, Y. et al., Sequential targeting of the genes encoding immunoglobulin-μ and prion protein in cattle, *Nat. Genet.* Online publication, June 6, 2004.
16. Stice, S.L. and Robl, J.M., Nuclear reprogramming in nuclear transplant rabbit embryos, *Biol. Reprod.*, 39, 657, 1988.
17. Collas, P. and Robl, J.M., Factors affecting the efficiency of nuclear transplantation in rabbit embryos, *Biol. Reprod.*, 43, 877, 1990.
18. Heyman, Y., Chesné, P. and Renard, J.P., Reprogrammation complète de noyaux embryonnaires congelés, après transfert nucléaire chez le lapin, *C.R. Acad. Sci. Paris,* t. 311, Série III, 321, 1990.
19. Yang, X.Y. et al., Nuclear totipotency of cultured rabbit morulae to support full-term development following nuclear transfer, *Biol. Reprod.*, 47, 636, 1992.
20. Piotrowska, K. et al., Effects of preactivation of ooplasts or synchronization of blastomere nuclei in G1 on preimplantation development of rabbit serial nuclear transfer embryos, *Biol. Reprod.*, 63(3), 677, 2000.
21. Fan, J. and Watanabe, T., Transgenic rabbits as therapeutic protein bioreactors and human disease models, *Pharmacol. Therapeut.*, 99, 261, 2003.
22. Chesné, P. et al., Cloned rabbit produced by nuclear transfer from adult somatic cells, *Nat. Biotechnol.*, 20, 366, 2002.
23. Challah–Jacques, M., Chesné, P. and Renard, J.P., Production of cloned rabbits by somatic nuclear transfer, *Cloning Stem Cells, Methods Paper*, 5(4), 295, 2003.
24. Chesné, P. et al., Production of rabbit clones for atherogenesis studies, in *15th A.E.T.E. Scientific Meeting*, Lyon, 1999, 136.
25. Wakayama, T. et al., Full-term development of mice from enucleated oocytes injected with cumulus cell nuclei, *Nature*, 394, 369, 1998
26. Wells, D.N., Misica, P.M. and Tervit, H.R., Production of cloned calves following nuclear transfer with cultured adult mural granulosa cells, *Biol. Reprod.*, 60, 996, 1999.
27. Larman, M.G. et al., Cell cycle-dependent Ca^{2+} oscillations in mouse embryos are regulated by nuclear targeting of PLCζ, *J. Cell Sci.*, 117, 2513, 2004.
28. Jones, K.T., Turning it on and off: M-phase promoting factor during meiotic maturation and fertilization, *Mol. Hum. Reprod.*, 10, 1, 2004.
29. Fan, H.Y. and Sun, Q.Y., Involvement of MAPK cascade during oocyte maturation and fertilization in mammals, *Biol. Reprod.*, 70, 535, 2004
30. McLay, D.W. and Clarke, H.J., Remodeling the paternal chromatin at fertilization in mammals, *Reproduction*, 125, 625, 2003.
31. Fulka, Jr., J. et al., Nucleus transfer in mammals: how the oocyte cytoplasm modifies the transferred nucleus, *Theriogenology*, 55, 1373, 2001.
32. Adenot, P.G. et al., Differential H4 acetylation of paternal and maternal chromatin precedes DNA replication and differential transcriptional activity in pronuclei of one-cell mouse embryos, *Development*, 124, 4615, 1997.
33. Ozil, J.P., The parthenogenetic development of rabbit oocytes after repetitive pulsatile electrical stimulation, *Development*, 109, 117, 1990.

34. Ozil, J.P. and Huneau, D., Activation of rabbit oocytes: the impact of the Ca^{2+} signal regime on development, *Development*, 128, 917, 2001.
35. Ducibella, T. et al., Egg-to-embryo transition is driven by differential responses to Ca(2+) oscillation number, *Dev. Biol.*, 250, 280, 2002.
36. Dinnyès, A. et al., Development of cloned embryos from adult rabbit fibroblasts: effect of activation treatment and donor cell preparation, *Biol. Reprod.*, 64, 257, 2001.
37. Escriba, M.J. and Garcia–Ximenez, F., Electroactivation of rabbit oocytes in an hypotonic pulsing medium and parthenogenetic *in vitro* development without cytochalasin B-diploidizing pretreatment, *Theriogenology*, 51, 963, 1999.
38. Escriba, M.J. and Garcia–Ximenez, F., Influence of sequence duration and number of electrical pulses upon rabbit oocyte activation and parthenogenetic *in vivo* development, *Anim. Reprod. Sci.*, 59, 99, 2000.
39. Collas, P. and Robl, J.M., Relationship between nuclear remodeling and development in nuclear transplant rabbit embryos, *Biol. Reprod.*, 45, 455, 1991.
40. Adenot, P.G. et al., *In vivo* aging of oocytes influences the behavior of nuclei transferred to enucleated rabbit oocytes, *Mol. Reprod. Dev.*, 46, 325, 1997.
41. Yin, X.J. et al., Development of rabbit parthenogenetic oocytes and nuclear-transferred oocytes receiving cultured cumulus cells, *Theriogenology,* 54, 1469, 2000.
42. Liu, C.T. et al., Parthenogenesis of rabbit oocytes activated by different stimuli, *Anim. Reprod. Sc.*, 70, 267, 2002.
43. Pinto–Correia, C. et al., Chromatin and microtubule organization in the first cell cycle in rabbit parthenotes and nuclear transplant embryos, *Mol. Reprod. Dev.*, 34, 33, 1993.
44. Mitalipov, S.M. et al., Development of nuclear transfer and parthenogenetic rabbit embryos activated with inositol 1,4,5-triphosphate, *Biol. Reprod.*, 60, 821, 1999.
45. Yin, X.J., Kato, Y. and Tsunoda, Y., Effect of enucleation procedures and maturation conditions on the development of nuclear-transferred rabbit oocytes receiving male fibroblast cells, *Reproduction*, 124, 41, 2002.
46. Liu, J.L. et al., Differential development of rabbit embryos derived from parthenogenesis and nuclear transfer, *Mol. Reprod. Dev.*, 68, 58, 2004.
47. Oprescu, S. and Thibault, C., Duplication de l'ADN dans les oeufs de lapine après la fécondation, *Ann. Biol. Anim. Bioch. Biophys.*, 5, 151, 1965.
48. Szöllösi, D., Time and duration of DNA synthesis in rabbit eggs after sperm penetration, *Anat. Rec.*, 154, 209, 1966.
49. Henrion, G. et al., Differential regulation of the translation and the stability of two maternal transcripts in preimplantation rabbit embryos, *Mol. Reprod. Dev.*, 56, 12, 2000.
50. Kanka, J. et al., Transcriptional activity and nucleolar ultrastructure of embryonic rabbit nuclei after transplantation to enucleated oocytes, *Mol. Reprod. Dev.*, 43, 135, 1996.
51. De Sousa, P.A. et al., Reprogramming of fibroblast nuclei after transfer into bovine oocytes, *Cloning*, 1, 63, 1999.
52. Daniels, R., Hall, V. and Trounson, A.O., Analysis of gene transcription in bovine nuclear transfer embryos reconstructed with granulosa cell nuclei, *Biol. Reprod.*, 63, 1034, 2000.
53. Wrenzycki, C. et al., Nuclear transfer protocol affects messenger RNA expression patterns in cloned bovine blastocysts, *Biol. Reprod.*, 65, 309, 2001.
54. Heyman, Y. et al., Frequency and occurrence of late-gestation losses from cattle cloned embryos, *Biol. Reprod.*, 66, 6, 2002.
55. Pacheco–Trigon, S. et al., Molecular characterization of genomic activities at the onset of zygotic transcription in mammals, *Biol. Reprod.*, 67, 1907, 2002.

56. Wade, P.A. and Kikyo, N., Chromatin remodeling in nuclear cloning, *Eur. J. Biochem.*, 269, 2284, 2002.
57. Dean, W. et al., Altered imprinted gene methylation and expression in completely ES cell-derived mouse fetuses: association with aberrant phenotypes, *Development,* 125(12), 2273, 1998.
58. Young, C.E. and Fairburn, H.R., Improving the safety of embryo technologies: possible role of genome imprinting, *Theriogenology,* 53(2), 627, 2000.
59. Khosla, S. et al., Culture of preimplantation mouse embryos affects fetal development and the expression of imprinted genes, *Biol. Reprod.*, 64, 918, 2001.
60. Oswald, J. et al., Active demethylation of the paternal genome in the mouse zygote, *Curr. Biol.*, 10(8), 475, 2000.
61. Dean, W. et al., Conservation of methylation reprogramming in mammalian development: aberrant reprogramming in cloned embryos, *Proc. Natl. Acad. Sci. USA*, 98, 13734, 2001.
62. Bourc'his, D. et al., Delayed and incomplete reprogramming of chromosome methylation patterns in bovine cloned embryos, *Curr. Biol.*, 11, 1542, 2001.
63. Beaujean, N. et al., Nonconservation of mammalian preimplantation methylation dynamics, *Curr. Biol.*, 14, 266, 2004.
64. Shi, W. et al., Methylation reprogramming and chromosomal aneuploidy in *in vivo* fertilized and cloned rabbit preimplantation embryos, *Biol. Reprod.,* 71, 340, 2004.
65. Christians, E., Rao, V.H. and Renard, J.P., Sequential acquisition of transcriptional control during early embryonic development in the rabbit, *Dev. Biol.*, 164(1), 160, 1994.
66. Wilmut, I. et al., Viable offspring derived from fetal and adult mammalian cells, *Nature*, 385, 810, 1997.
67. Kato, Y. et al., Eight calves cloned from somatic cells of a single adult, *Science*, 282, 2095, 1998.
68. Baguisi, A. et al., Production of goats by somatic cell nuclear transfer, *Nat. Biotechnol.*, 17(5), 456, 1999.
69. Kubota, C. et al., Six cloned calves produced from adult fibroblast cells after long-term culture, *Proc. Natl. Acad. Sci. USA*, 97, 990, 2000.
70. Vignon, X. et al., Production of bovine embryos by nuclear transfer of adult skin fibroblasts: the effect of serum starvation, *Theriogenelogy*, 53, 245, 2000.
71. Hill, J.R. et al., Development rates of male bovine nuclear transfer embryos derived from adult and fetal cells, *Biol. Reprod.*, 62, 1135, 2000.
72. Cibelli, J.B. et al., Cloned transgenic calves produced from nonquiescent fetal fibroblasts, *Science*, 280, 1256, 1998.
73. Kühholzer, B. et al., Clonal lines of transgenic fibroblast cells derived from the same fetus result in different development when used for nuclear transfer in pigs, *Biol. Reprod.*, 64, 1695, 2001.
74. Ono, Y. et al., Cloned mice from fetal fibroblast cells arrested at metaphase by a serial nuclear transfer, *Biol. Reprod.*, 64, 44, 2001.
75. Kasinathan, P. et al., Production of calves from G1 fibroblasts, *Nat. Biotechnol.*, 19, 1176, 2001.
76. Collas, P., Balise, J.J. and Robl, J.M., Influence of cell cycle stage of the donor nucleus on development of nuclear transplant rabbit embryos, *Biol. Reprod.*, 46, 492, 1992.
77. Collas, P. et al., Effects of donor cell cycle stage on chromatin and spindle morphology in nuclear transplant rabbit embryos, *Biol. Reprod.*, 46, 501, 1992.

78. Cervera, R.P. and Garcia–Ximenez, F., Oocyte age and nuclear donor cell type affect the technical efficiency of somatic cloning in rabbits, *Zygote*, 11, 151, 2003.
79. Polajaeva, I.A. et al., Cloned pigs produced by nuclear transfer from adult somatic cells, *Nature*, 407, 86, 2000.
80. Kato, Y., Tani, T. and Tsunoda, Y., Cloning of calves from various somatic cell types of male and female adult, newborn and fetal cows, *J. Reprod. Fertil.*, 120, 231, 2000.
81. Forsberg, E.J. et al., Production of cloned cattle from *in vitro* systems, *Biol. Reprod.*, 67(1), 327, 2002.
82. Lagutina, I.S. et al., Developmental potential of rabbit nuclear transfer embryos produced by various fusion/activation protocols, *Theriogenology*, 53, 230 abstr., 2000.
83. Colman, A. and Kind, A., Therapeutic cloning: concepts and practicalities, *Trends Biotechnol.*, 18(5), 192, 2000.
84. Rideout III, W.M., Eggan, K. and Jaenish, R., Nuclear cloning and epigenetic reprogramming of the genome, *Science*, 293, 1093, 2001.
85. Du, F. et al., Nuclear transfer of putative rabbit embryonic stem cells leads to normal blastocyst development. *J. Reprod. Fert.* 104, 219, 1995.
86. Moens, A. et al., Differential ability of male and female rabbit fetal germ cell nuclei to be reprogrammed by nuclear transfer, *Differentiation*, 60(5), 339, 1996.
87. Humpherys, D. et al., Epigenetic instability in ES cells and cloned mice, *Science*, 293, 95, 2001.
88. Zhou, Q. et al., Development potential of mouse embryos reconstructed from metaphase embryonic stem cell nuclei, *Biol. Reprod.*, 65, 412, 2001.
89. Baqir, S. and Smith, L.C., Growth restricted *in vitro* culture conditions alter the imprinted gene expression patterns of mouse embryonic stem cells, *Cloning Stem Cells*, 5, 199, 2003.
90. Gao, S. et al., Somatic cell-like features of cloned mouse embryos prepared with cultured myoblast nuclei, *Biol. Reprod.*, 69, 48, 2003.
91. Brunet–Simon, A. et al., Onset of zygotic transcription and maternal transcript legacy in the rabbit embryo, *Mol. Reprod. Dev.*, 58, 127, 2001.
92. Chang, M.C., Development and fate of transferred rabbit ova or blastocysts in relation to the ovulation time of recipients, *J. Exp. Zool.*, 114, 197, 1950.
93. Tsunoda, Y., Soma, T. and Sugie, T., Effect of postovulatory age of recipient on survival of frozen–thawed rabbit morulae, *J. Reprod. Fert.*, 65, 483, 1982.
94. Greenwald, G.S., The role of the mucin layer in development of the rabbit blastocyst, *Anat. Rec.*, 142, 407, 1962.
95. Murakami, H. and Imai, H., Successful implantation of *in vitro* culture rabbit embryos after uterine transfer: a role for mucin, *Mol. Reprod. Dev.*, 43, 167, 1996.
96. Walton, A. and Hammond, J., Observations in ovulation in the rabbit, *J. Exp. Biol.*, 6, 190, 1929.
97. Heape, W., Preliminary note of the transplantation and growth of mammalian ova within a uterine foster mother, *Proc. R. Soc. Lond. A/B,* 48, 457, 1890.
98. Boiani, M. et al., Pluripotency deficit in clones overcome by clone–clone aggregation: epigenetic complementation?, *EMBO J.*, 22, 5304, 2003.
99. McLaren, A., Sex chimerism and germ cell distribution in a series of chimeric mice, *J. Embryol. Exp. Morphol.*, 33, 205, 1975.
100. Williams, T.J., Munro, R.K. and Shelton, J.N., Production of interspecies chimeric calves by aggregation of Bos taurus and Bos indicus demi-embryos, *Reprod. Fertil. Dev.*, 2(4), 385, 1990.
101. Boediono, A. et al., Offspring born from chimeras reconstructed from parthenogenetic and *in vitro* fertilized bovine embryos, *Mol. Reprod. Dev.*, 53, 159, 1999.

102. Babinet, C. and Bordenave, G.R., Chimeric rabbits from immunosurgically prepared inner-cell-mass transplantation, *J. Embryol. Exp. Morph.*, 60, 429, 1980.
103. Moens, A. et al., Low levels of chimerism in rabbit fetuses produced from preimplatation embryos microinjected with fetal gonadal cells, *Mol. Reprod. Dev.*, 43, 38, 1996.
104. Galat, V. et al., Effect of donor cell age on the efficiency of nuclear transfer in rabbits, *Reprod. Biomed Online*, 4(1), 32, 2002.
105. Inoue, K. et al., Improved postimplantation development of rabbit nuclear transfer embryos by activation with inositol 1,4,5-triphosphate, *Cloning Stem Cells*, 4(4), 311, 2002.
106. Li, G.P. et al., Rabbit cloning: improved fusion rates using cytochalasin B in the fusion buffer, *Mol. Reprod. Dev.*, 61, 187, 2002.
107. Nowshari, M.A. et al., Nuclear transfer with adult and follicular cells in rabbits, *Theriogenology*, 57(1), 439, 2002.
108. Pitts, J.D. and Burk, R.R., Specificity of junctional communication between animal cells, *Nature*, 267, 762, 1976.
109. Ducibella, T. et al., The preimplantation mammalian embryo: characterization of intracellular junctions and their appearance during development, *Dev. Biol.*, 45, 231, 1975.
110. Yin, X.J., Kato, Y. and Tsunoda, Y., Effect of delayed enucleation on the developmental potential of nuclear-transferred oocytes receiving adult and fetal fibroblast cells, *Zygote*, 10(3), 217, 2002.
111. Matsuda, J. et al., Production of transgenic chimera rabbit fetuses using somatic cell nuclear transfer, *Cloning Stem Cell*, 4(1), 9, 2002.
112. Zhou, Q., Boulanger, L. and Renard, J.P., A simplified method for the reconstruction of fully competent mouse zygotes from adult somatic donor nuclei, *Cloning*, 2(1), 35, 2000.
113. Pincus, G. and Werthessen, N.T., The comparative behavior of mammalian eggs *in vivo* and *in vitro*, *J. Exp. Zool.*, 78, 1, 1936.
114. Chantry–Darmon, C. et al., 133 new gene localizations on the rabbit cytogenetic map, *Cytogenet. Genome Res.*, 103, 192, 2003.

7 Evidence For and Against Associations between ART and Congenital Malformation Syndromes

Michael R. DeBaun, Aimee S. Chang, Michael Wangler, and Kelle H. Moley

CONTENTS

7.1 INTRODUCTION

Beckwith–Wiedemann syndrome (BWS) is a congenital disorder associated with various findings, including macrosomia, macroglossia, midline abdominal defects, and a predisposition to embryonal cancer (MIM 130650). Genetic mutations [1] and a loss of imprinting (LOI) of a group of imprinted genes occurring on 11p15 are involved with BWS [2]. The epigenetic changes include aberrant methylation and imprinting of LIT1 occurring in 42% of patients (MIM 604115) [1,3]; H19 in 11% of patients (MIM 103280) [1]; or a combination of these two in 13% of patients in the setting of uniparental disomy (75%) or in conjunction with an extensive imprinting disturbance throughout 11p (25%) [1].

7.2 ASSOCIATIONS CONCERNING BWS, ANGELMAN SYNDROME, AND ART

Recently, an association between assisted reproductive technology (ART) and congenital malformation syndromes was documented [4,5]. Three groups have reported an increased rate of ART conceptions among children with BWS [4,6,7]. Taken together these results suggest that approximately 4% of the children with BWS were born after ART [4,6,7].

Imprinting abnormalities are also associated with Angelman syndrome, which is characterized by mental retardation, motor defects, and lack of speech; recent studies have implicated ART in some cases of Angelman syndrome (MIM 105830) [5]. In this syndrome, children are born with imprinting abnormalities of a gene cluster on chromosome 15, resulting in a loss of function of the maternal allele of Ubiquitin-protein ligase E3A (UBE3A) (MIM 601623) [8]. In three cases, children with Angelman syndrome were conceived after *in vitro* fertilization (IVF) with intracytoplasmic sperm injection (ICSI). This gives further support for the association between congenital malformation syndromes with imprinting mutations and ART [5,9].

7.3 CULTURE MEDIUM DOES NOT CAUSE BWS TO OCCUR AFTER ART

Based on work in animal models, we initially hypothesized that the culture medium used for the embryos or blastocysts could be a common factor among children with BWS born after IVF. Thompson et al. demonstrated that they could decrease the frequency of large offspring syndrome (LOS) in farm animals following ART by changing the protein supplementation in the culture medium from sera to albumin and amino acids [10]. LOS refers to a syndrome that occurs in sheep and cows developed through ART [11]. Also, Doherty et al. showed that, in preimplantation mouse embryos, the type of culture medium could alter the expression of H19 [12].

To test the hypothesis that the culture is the cause of BWS after ART, a review was conducted of laboratory data from the reproductive endocrine centers of mothers in the BWS Registry who gave birth to children with BWS following ART. Parents of children born with BWS were asked to fill out a questionnaire that specifically asked whether ART was used. If ART was utilized, then the type of ART was ascertained (i.e., IVF, IVF with ICSI, gonadotropin stimulation with IUI, or clomiphene citrate stimulation with natural intercourse). The medical records from the Reproductive Endocrinology and Infertility Centers were then obtained for further review.

Nineteen patients were found whose mothers had undergone some type of ART. The reproductive endocrine medical records of 12 of 19 mothers were obtained, and each medical record contained extensive information on ART treatment and management course. Interestingly, all 12 mothers were treated at different reproductive endocrine and infertility centers. Ten out of twelve of the patients underwent IVF. Of the 10 mothers who received IVF, 50% underwent ICSI. One of the five mothers had a partial ICSI procedure with half of the embryos inseminated by traditional insemination and the other half inseminated by ICSI. In five sets of dizygotic twins among the women who had IVF, only one twin was affected. In the two patients who were not conceived through IVF, one child was conceived through clomiphene citrate stimulation and artificial insemination of donor sperm, and another patient was conceived through gonadotropin stimulation with intrauterine insemination. Table 7.1 describes the method of ART in the 12 children.

All twelve women received some type of ovarian stimulatory medication. Two mothers received medications alone (clomiphene citrate or gonadotropins) and ten mothers received ovarian stimulation as part of their IVF treatment. The quantity of gonadotropins administered for stimulation ranged from 16 to 102 ampoules for the ten women who received IVF.

The number and timing of embryos transferred also varied (Table 7.2). The number of embryos/blastocysts transferred varied from one to eight. Five of the patients had embryo transfers three days postfertilization; the other five had blastocysts transferred on day 5.

TABLE 7.1
Sources for Comparison of Features of LOS Resulting from ART to Beckwith–Wiedemann Syndrome

Feature	LOS Resulting from ART	Patients with BWS
Hypoglycemia	[13]	[1,14,15]
Macrosomia	Cattle [16,17] Sheep [18,19]	[1,14,15]
Organomegaly	Cattle [20] Sheep [23,24]	[21,22]
Polyhydramnios	Sheep [24]	[14,25,26]
Placentomegaly	[27]	[28–30]
Altered IGF-II molecular signaling	[31]	[1,4]

TABLE 7.2
Associated Reproductive Techniques Used before Conception in Children with BWS Born after ART

Patient	Day of Transfer	Use of ICSI	Culture Media
Patient 1	3	No	P1 +SSS
Patient 2	5	Yes	G1, G2
Patient 3	5	Yes	G1, G2
Patient 4	5	No	HTF +SSS
Patient 5	5	No	P1 blastocyst media
Patient 6	3	Partial	HTF +SSS
Patient 7	3	Yes	P1 +SSS, HTF
Patient 8	3	No	P1
Patient 9	3	Yes	Cook's sequential
Patient 10	5	No	G1, G2
	Assisted Reproduction Without IVF		
Patient 11	Clomiphene citrate and intrauterine insemination		
Patient 12	Ovulation induction and intrauterine insemination		

Similarly, great variation was present in the media, and no common media were discovered. Five different types of media were used for embryo culture. The media included (Table 7.2):

- Sequential media from G1 to G2 (*in vitro* life) for blastocysts ($n = 3$)
- Human tubal fluid or HTF (Irvine media) + serum substitute supplement (SSS) ($n = 2$)
- P1 ($n = 1$)
- P1 + SSS ($n = 1$)
- Cook's sequential media ($n = 1$)
- Sequential media of P1 + SSS and then HTF ($n = 1$)
- Sequential culturing in HTF, then P1 media, and then blastocyst media + SSS ($n = 1$)

G1, G2, blastocyst medium, and Cook's sequential medium contained a number of amino acids; HTF and P1 had no amino acid constituents. Each of the media combinations had a human sera albumin component as a supplement (SSS) or within the media (G1,G2, and Cook's sequential media) except for one case of P1 alone.

The only common factor discovered so far is that all 12 women received some type of ovarian stimulatory medications. Because the study group is small, these findings and the conclusions that can be drawn are limited. Another possible hypothesis that has not been examined may be that mothers of children with BWS have an underlying infertility problem that may be associated with the disease. However, this is unlikely because none of the mothers had features characteristic of

BWS. At this point, larger prospective studies need to be initiated to learn whether an association between ART and BWS truly exists.

7.4 LOS AND BWS

The association between Beckwith–Wiedemann syndrome and assisted reproductive technology receives indirect support from a phenomenon observed in cattle and sheep known as the large offspring syndrome (LOS). Like BWS, LOS is associated with reproduction technology — specifically, *in vitro* embryo production (IVP). This section describes the significant phenotypic and molecular overlap between BWS and LOS.

7.4.1 Macrosomia

BWS is one of several congenital syndromes associated with macrosomia. However, in BWS the macrosomia occurs primarily prenatally. Defined as birth weight above the 90th percentile, macrosomia was present in 65% of cases in the BWS registry (95% CI 60–70) and 58% of children with BWS resulting from ART (95% CI 36–77) (M. DeBaun, BWS Registry, unpublished data). Likewise, the most distinctive feature of LOS is the increase in birth weight.

Several studies on calves have compared the birth weights of animals derived from IVP and compared them to those of calves derived from artificial insemination or natural mating [17,27,32–36]. Increased birth weight was initially thought to be the sole abnormality found in calves produced *in vitro* [35]. Since then, several groups have confirmed an increase in the average birth weight among groups of calves derived from IVP compared to control groups derived from artificial insemination or natural mating [32,33,35,36]. An increase in mean birth weight between these groups appears to result from an overgrowth phenotype that affects only a portion of the calves derived from *in vitro* production [17,32]. This suggests that, like BWS, LOS appears in a fraction of the individuals undergoing *in vitro* procedures.

Similarly, IVP of ovine embryos has also led to increased birth weights among sheep [10,19,23,37–39]. The increased average birth weight in sheep derived from IVP was the first abnormality noted to be associated with *in vitro* procedures [38]. As in bovines, only a proportion of ovine offspring derived from IVP appear to be macrosomic [38]. Not only are the effects on birth weight observed in a subset of lambs, but individuals within groups appear variably affected; particular lambs five times the average birth weight [38] have been reported.

Macrosomia, the cardinal feature of BWS and LOS, is common after IVF and IVP, respectively. The pathogenetic mechanisms underlying macrosomia in BWS and LOS are unknown.

7.4.2 Organomegaly

In addition to macrosomia, patients with BWS have organomegaly — specifically, heart, liver, kidney, and tongue [14,22]. Similarly, the organ weights of large farm animals derived from IVP have been shown to be higher than those in control animals, a finding initially attributed to the presence of macrosomia in animals with LOS [20]. Enlarged liver [23,34]; heart [20,23]; kidney [23]; testicles [39]; and portions of the hind leg [23,34]; have been observed. The use of allometric growth coefficients allowed an evaluation of organ growth in macrosomic individuals to determine whether the organ enlargement is beyond that to be expected in macrosomic individuals [23,24]. Although growth out of proportion to the body size has been clearly demonstrated [40], organomegaly is not present in all macrosomic animals in LOS [37].

7.4.3 Polyhydramnios

Polyhydramnios, increased fluid in the amniotic sac, has been identified in a number of cases of BWS and was found in 40% (95%CI 35–46) of the patients in the BWS Registry and 74% (95%CI 51–88) of children with BWS conceived after ART [26]. Polyhydramnios is one of the most common abnormalities observed during a pregnancy that results in a child with BWS.

Polyhydramnios has also been reported during pregnancies from ovine embryos produced *in vitro* [40], as well as bovine pregnancies [16,17] under similar conditions. Whether all or most of these pregnancies resulted in calves with high birth weights or other features of LOS has not been clarified. Nonetheless, polyhydramnios appears to be a common feature of BWS with and without ART as well as following IVP.

7.4.4 Abnormalities of the Placenta

The placenta is abnormal and enlarged in BWS. A number of characteristic lesions of the placenta have been reported to be significantly associated with BWS. In at least some proportion of cases of BWS, the placenta is characterized by villous hydrops without trophoblast hyperplasia. This is an unusual histopathologic finding in the placenta because villous hydrops is almost always found associated with trophoblast hyperplasia in the setting of hydatidiform mole and isolated villous hydrops [28]. In addition, the presence of mesenchymal dysplasia, an increase in the connective tissue between the fetal circulation and the trophoblast, has also been found in a significant proportion of cases of BWS [29,30].

The finding of hydrallantois in LOS suggests that, as in BWS, LOS may be characterized by abnormalities of the placenta [16]. However, no formal microscopic examination of the placenta to determine whether findings such as villous hydrops or mesenchymal dysplasia are present in the placenta of LOS has been reported. The only reported abnormalities of the placenta in LOS have been that of a smaller number of normal placentomes, a unit surrounding the spiral arteries, in the placentas of animals with LOS [20]. The significance of this finding and whether it might relate to histopathologic placental abnormalities are unclear.

The specific characteristics of the placenta in LOS are unknown; however, the size of the placenta is enlarged in LOS. Some groups have proposed that the spectrum of overgrowth and physiologic abnormalities of LOS may result from perturbations in placental function [13,27]. This hypothesis is supported by a positive correlation between fetal and placental weights in LOS [24]; however, separate studies have suggested cases of LOS in which no correlation exists [23,37].

Irrespective of whether the placenta abnormalities can account for the overgrowth phenotypes in LOS, abnormal placental phenotypes are observed in LOS and BWS. Placentomegaly occurs in both syndromes, although it is unclear whether the causes of the placentomegaly are equivalent. Histopathologic and functional placental studies will need to be performed before conclusions can be drawn.

7.4.5 Molecular Alterations in IGF II Signaling

The molecular etiology of BWS is related to abnormal methylation or imprinting of genes on chromosome 11p15. A number of different genetic and epigenetic molecular abnormalities have been identified in patients with BWS. Approximately 42% of BWS patients have aberrant methylation and imprinting of LIT1, an untranslated RNA in the KvLQT1 gene [1,3]. Approximately 11% have aberrant methylation of H19 [1], an untranslated RNA associated with abnormal expression of insulin-like growth factor 2(IGFII). Approximately 13% of patients with BWS have altered methylation of H19 and LIT1. Of these, 75% have uniparental disomy (UPD) of 11p15 and the remaining 25% have an extensive imprinting disturbance throughout 11p15 [1,2]. Finally, 5% of BWS patients have a mutation in $P57^{K1P2}$ [41]. In children without BWS born after ART, abnormal methylation of HI9 and LITI has been noted. Abnormality methylation of HI9 is associated with loss of imprinting of IGF2 [1].

Similarly, in large farm animals born after IVP, abnormal imprinted genes associated with growth have been identified — specifically, IGF2 receptor. A causative role for IGFII has been proposed in LOS [42]. This hypothesis has been tested in LOS in sheep [31]. In a group of 12 sheep born after IVP, 9 sheep appeared to share loss of methylation at a differentially methylated region of the IGFII receptor gene.

7.4.6 Disparate Features

Although BWS and LOS share similar features, a number of features appear to differ. Much of the clinical significance of BWS originates from the increased incidence of embryonal tumors [15]. Some evidence suggests that LOS may be characterized by embryonal tumors; however, the incidence is low and highly variable (K. Sinclair, personal communication). In addition, a number of the congenital malformations characterizing BWS, such as abdominal wall defects, ear defects, and hemihypertrophy, have not been reported in LOS.

Parturition is clearly affected in BWS and LOS, but with opposite effects. Children in the BWS Registry were delivered preterm at a substantially higher rate (40%) than the general population [26]. The higher rates of polyhydramnios and preeclampsia appear to contribute to the higher rate of preterm births. However, a substantial fraction (12%) of these pregnancies had preterm labor requiring hospitalization and tocolysis [26]. In contrast, in cattle [32] and sheep [38], LOS appears to be characterized by a failure to go into labor at term.

The association of an overgrowth phenotype with ART in humans is supported indirectly by the fact that *in vitro* production of cattle and sheep also results in an overgrowth syndrome. BWS and LOS share similar phenotypic features, such as macrosomia, organomegaly, placentomegaly, and polyhydramnios. The IGF II signaling pathway has been implicated in BWS and in LOS. The common features of LOS and BWS are shown in Table 7.1.

7.5 ENVIRONMENTAL EFFECTS OF ART ON GENOMIC IMPRINTING: EVIDENCE IN NONHUMAN MAMMALS

The *in vitro* culture condition in which the zygote undergoes early cleavage is the most likely factor common in all types of ART that could account for changes in genomic imprinting. Differences in embryonic gene expression between *in vivo* and *in vitro* cultured murine embryos were first reported by Christians et al. [43]. Luciferase reporter gene activity was 15-fold greater in embryos cultured *in vitro* than that seen in the embryos derived *in vivo*. Khosla et al. [44] showed that culture of murine embryos in fetal calf serum (FCS) adversely affected imprinted gene expression and subsequent development as compared to embryos cultured without FCS. H19 and IGF2 expression were decreased, whereas Grb10 expression was increased, due to changes in genomic methylation. In addition, embryos cultured in FCS experienced decreased survival to day 14 and abnormal growth in fetuses that survived as compared to controls cultured without FCS.

Similarly, Doherty et al. [12] demonstrated that embryos cultured in the high glucose containing Whitten's media aberrantly expressed the imprinted gene H19 due to a loss in methylation. Recently, Lane and Gardner [45] have reported that ammonia generated by breakdown of amino acids present in media when cultured for prolonged periods at 37°C has similar adverse effects on imprinted gene expression. They demonstrated once again that high concentrations of ammonia induced abnormal methylation of H19, resulting in metabolic and intracellular pH changes as well as increased apoptosis at the blastocyst stage. The fate of these embryos when transferred also differed from controls. Fewer blastocysts implanted and, of those that did, fetal growth was significantly blunted.

Gene expression in the preimplantation embryos is also influenced by ion concentrations. Changes in expression of IGF-1, IGF-II, and IGF-I receptors were reported in embryos cultured in a simplex optimized medium containing 85 m*M* NaClI-1 as compared to those containing 125 m*M* NaClI-1. The higher ion concentration led to reduced RNA synthesis rates.

Data from studies of assisted reproductive technologies in cattle and sheep also report aberrant fetal growth and alterations in imprinted genes [31]. With nuclear transfer (NT)-derived and *in vitro*-produced (IVP) embryos, *in vitro* culture for extended periods of time in sera has been attributed in part to this epigenetic phenomenon; however, the definitive mechanisms have yet to be determined.

Epigenetic consequences appear to be common among species although; the phenotypic effects on the fetus differ [10]. Specifically, the effect of perturbed epigenetic modifications on growth is drastically different in mice and ruminants. In mouse, the effects of prolonged *in vitro* culture are reduced fetal weight and fetal death. Cloning or NT in mice results in increased placental weight and fetal or perinatal death. In ruminants, IVP embryos experience higher birth weights and skeletal and organ abnormalities, referred to as LOS. Similarly, cloning in these species also results in higher birth weight and fetal death.

These data would strongly suggest that environmental stress induced by *in vitro* manipulation of fertilization and early embryo cleavage stages has been associated with the epigenetic regulation of genomic imprinting and the abnormal fetal growth and development in experimental animals. Animal studies show that these stresses may be induced by chemically defined media containing protein or sera. How this occurs is less clear. It is possible that altered cleavage rates due to *in vitro* conditions may also predispose the embryos to these epigenetic alterations [44].

DNA methyltransferases (DNMTs) and thus DNA methylation are highly regulated in mouse and, as most recently shown, in human oocytes and preimplantation embryos. In mouse, DNMTs regulate maternal genomic imprinting during oogenesis and early embryogenesis. Environmental changes in expression of any of the DNMTs would lead to aberrant imprinting. Changes in chromatin remodeling and histone modifications may also be responsible. Recent studies in mouse oocytes have demonstrated the importance of the vertebrate insulator protein, CCCTC binding factor (CTCF), in the maintenance of H19 maternal allele-specific hypomethylation [46].

Environmental conditions may alter important interaction of these epigenetic modifiers, leading to abnormal methylation patterns. Abnormal development and methylation were demonstrated in embryos experiencing transgenic RNAi-induced decrease in CTCF expression. Finally, changes in histone deacetylation have been linked to DNA methylation and decreased gene expression. Embryos cultured *in vitro* are known to undergo significant metabolic alterations, thus this may be a mechanism by which such alterations could lead indirectly to the epigenetic medications.

Finally, how environmental stresses induced by ART include epigenetic modifications in nonhuman mammalian embryos is still an active area of research. Whether the same conclusions can be applied to humans undergoing ART is still unknown.

7.6 SUMMARY

Based on the evidence provided thus far, arguments for and against the association between BWS and IVF can be made. Observations in favor of the association between BWS and IVF include multiple studies from several countries (three in total) demonstrating that children with BWS born after IVF occur at a greater rate than would be expected by chance. In addition, multiple corollaries between BWS and ruminants born after IVF include the presence of macrosomia, hypoglycemia, placenta abnormalities, and organomegaly. Also, patients with BWS have been noted to have imprinting abnormalities of insulin like growth factor II and animals with LOS have been demonstrated to have imprinting defects in insulin like growth factor 2 receptor.

Similar to the LOS model, no one etiologic agent can be identified as being positive for the association between BWS and IVF. Finally, biologically plausible hypotheses, such as alterations in the rate of cleavage or chemical stressors in the media with some preliminary evidence from ruminant and murine models, indicate that expression or function of proteins (such as DNMTs and CTCF) that regulate imprinting in the early embryo could be perturbed by one or more factors during the process of IVF. To date, specific factors that are shared in common by patients with BWS conceived with ART — other than ovarian stimulation — have not been identified. Clearly, elucidating the cause of this association will require prospective studies on bovine, ovine, and human subjects, as well as continuing progress in the basic biology underlying imprints and growth regulation.

In contrast, other lines of evidence weaken the argument for the association between BWS and ART. Specifically, no causal mechanism has been identified. In addition, only one prospective epidemiologic study demonstrating the statistical association between BWS and IVF has been conducted. The possibility exists that parents of children with BWS and infertility problems may seek medical attention more often compared to parents of children with BWS who do not have infertility problems. Such bias would result in an overestimate of the frequency of IVF among children with BWS.

Yet another not fully explored possibility is that parents of children with BWS born after ART may have a subtle form of BWS. Although this is a possibility, it seems slightly less likely because none of the parents of the children born after BWS had clinical features that could be associated with BWS. Only a study that addresses the genotype of the parents who had children born after IVF with BWS could determine this possibility.

When the evidence for and against the association for BWS and IVF is fully explored, the balance of evidence would strongly suggest that BWS is associated with IVF. Further investigation is warranted along with the development of rigorous methodology in the field of IVF, accompanied with animal models to elucidate the underlying mechanism for these associations.

REFERENCES

1. DeBaun, M.R. et al., Epigenetic alterations of H19 and LIT1 distinguish patients with Beckwith–Wiedemann syndrome with cancer and birth defects, *Am. J. Hum. Genet.*, 70(3), 604, 2002.
2. Weksberg, R. et al., Disruption of insulin-like growth factor 2 imprinting in Beckwith–Wiedemann syndrome, *Nat. Genet.*, 5, 143, 1993.
3. Lee, M.P. et al., Loss of imprinting of a paternally expressed transcript, with antisense orientation to KVLQT1, occurs frequently in Beckwith–Wiedemann syndrome and is independent of insulin-like growth factor II imprinting, *Proc. Natl. Acad. Sci. USA*, 96(9), 5203, 1999.
4. DeBaun, M.R. et al., Association of *in vitro* fertilization with Beckwith–Wiedemann syndrome and epigenetic alterations of LIT1 and H19, *Am. J. Med. Genet.*, 72, 156, 2003.

5. Cox, G.F. et al., Intracytoplasmic sperm injection may increase the risk of imprinting defects, *Am. J. Hum. Genet.*, 71(1), 162, 2002.
6. Gicquel, C.V. et al., *In vitro* fertilization may increase the risk of Beckwith–Wiedemann syndrome related to the abnormal imprinting of the KCN1OT gene, *Am. J. Hum. Genet.*, 72(5), 1338, 2003.
7. Maher, E.R. et al., Beckwith–Wiedemann syndrome and assisted reproduction technology (ART), *J. Med. Genet.*, 40(1), 62, 2003.
8. Nicholls, R.D. et al., Imprinting in Prader–Willi and Angelman syndromes, *Trends Genet.*, 14(5), 194, 1998.
9. Orstavik, K.H. et al., Another case of imprinting defect in a girl with Angelman syndrome who was conceived by intracytoplasmic semen injection, *Am. J. Hum. Genet.*, 72(1), 218, 2003.
10. Thompson, J.G. et al., Lamb birth weight is affected by culture system utilized during *in vitro* pre-elongation development of ovine embryos, *Biol. Reprod.*, 53(6), 1385, 1995.
11. Young, L.E. et al., Large offspring syndrome in cattle and sheep, *Rev. Reprod.*, 3(3), 155, 1998.
12. Doherty, A.S. et al., Differential effects of culture on imprinted H19 expression in the preimplantation mouse embryo, *Biol. Reprod.*, 62(6), 1526, 2000.
13. Garry, F.B. et al., Postnatal characteristics of calves produced by nuclear transfer cloning, *Theriogenology,* 45(1), 141, 1996.
14. Filippi, G. and McKusick, V.A., The Beckwith–Wiedmann syndrome, *Medicine* (Baltimore), 49(4), 279, 1970.
15. DeBaun, M.R. and Tucker, M.A., Risk of cancer during the first four years of life in children from the Beckwith–Wiedemann syndrome registry, *J. Pediatr.*, 132, 398, 1998.
16. Hasler, J.F. et al., Production, freezing and transfer of bovine IVF Embryos and subsequent calving results, *Theriogenology,* 43(1), 141, 1995.
17. Kruip, T. and den Daas, J., *In vitro* produced and cloned embryos: effects on pregnancy, parturition and offspring, *Theriogenology,* 47(1), 43, 1997.
18. Walker, S.K. et al., The production of unusually large offspring following embryo manipulation: concepts and challenges, *Theriogenology,* 45(1), 111, 1996.
19. Young, L.E. et al., Increased ovine fetal weight following transient asynchronous embryo transfer is not associated with increased placental weight at day 21 of gestation, *Theriogenology,* 45(1), 231, 1996.
20. Farin, P.W. and Farin, C.E., Transfer of bovine embryos produced *in vivo* or *in vitro*: survival and fetal development, *Biol. Reprod.*, 52(3), 676, 1995.
21. DeBaun, M.R. et al., Nephromegaly in infancy and early childhood: a risk factor for Wilms tumor in Beckwith–Wiedemann syndrome, *J. Pediatr.*, 132(3 Pt 1), 401, 1998.
22. Choyke P.L. et al., Nonmalignant renal disease in pediatric patients with Beckwith–Wiedemann syndrome, *AJR Am. J. Roentgenol.*, 171(3), 733, 1998.
23. Sinclair, K.D. et al., Culture of sheep zygotes can alter fetal growth and development, *Theriogenology,* 47, 380, 1997.
24. Sinclair, K.D. et al., Fetal growth and development following temporary exposure of day 3 ovine embryos to an advanced uterine environment, *Reprod. Fertil. Dev.,* 10(3), 263, 1998a.
25. Ranzini, A.C. et al., Intrauterine growth and ultrasound findings in fetuses with Beckwith–Wiedemann syndrome, *Obstet. Gynecol.*, 89(4), 538, 1997.

26. Wangler, M.F. et al., Factors associated with preterm delivery in mothers of children with Beckwith–Wiedemann syndrome: A case cohort study from the BWS registry, *Am. J. Med. Genet.*, 134(2), 187, 2005.
27. Bertolini, M. and Anderson, G.B., The placenta as a contributor to production of large calves, *Theriogenology,* 57(1), 181, 2002.
28. Lage, J.M., Placentomegaly with massive hydrops of placental stem villi, diploid DNA content, and fetal omphaloceles: possible association with Beckwith–Wiedemann syndrome, *Hum. Pathol.*, 22(6), 591, 1991.
29. Jauniaux, E. et al., Perinatal features associated with placental mesenchymal dysplasia, *Placenta,* 18(8), 701, 1997.
30. Ohyama, M. et al., Mesenchymal dysplasia of the placenta, *Pathol. Int.*, 50(9), 759, 2000.
31. Young, L.E. et al., Epigenetic change in IGF2R is associated with fetal overgrowth after sheep embryo culture, *Nat. Genet.*, 27(2), 153, 2001.
32. Sakaguchi, M. et al., Embryonic and calving losses in bovine mixed-breed twins induced by transfer of in vitro-produced embryos to bred recipients, *Anim. Reprod. Sci.*, 72(3–4), 209, 2002.
33. McEvoy, T.G. et al., Postnatal growth and development of Simmental calves derived from *in vivo* or *in vitro* embryos, *Reprod. Fertil. Dev.*, 10(6), 459, 1998.
34. Jacobsen, H.M. et al., Body dimensions and birth and organ weights of calves derived from *in vitro* produced embryos cultured with or without serum and oviduct epithelium cells, *Theriogenology,* 53(9), 1761, 2000.
35. Wilson, J.M. et al., Comparison of birth-weight and growth characteristics of bovine calves produced by nuclear transfer (cloning), embryo transfer and natural mating, *Anim. Reprod. Sci.,* 38(1–2), 73, 1995.
36. Numabe, T. et al., Birth weight and gestation length of Japanese black calves following transfer of embryos produced *in vitro* with or without co-culture, *J. Vet. Med. Sci.*, 63(5), 515, 2001.
37. Sinclair, K.D. et al., Conceptus growth and development following *in vitro* culture of ovine embryos in media supplemented with bovine sera, *Theriogenology,* 49(1), 218, 1998b.
38. Walker, S.K. et al., *In vitro* culture of sheep embryos without coculture — successes and perspectives, *Theriogenology,* 37(1), 111, 1992.
39. Brown, B.W. and Radziewic, T., *In vitro* production of sheep embryos and resultant lambs. Animal reproduction: research and practice, *Proc. 13th Int. Congr. Anim. Reprod.*, Sydney, Australia, 3, P22–16, 1996.
40. Sinclair, K.D. et al., Aberrant fetal growth and development after *in vitro* culture of sheep zygotes, *J. Reprod. Fertil.*, 116(1), 177, 1999.
41. Lee, M.P. et al., Low frequency of p57kip2 mutation in Beckwith–Wiedemann syndrome, *Am. J. Hum. Genet.*, 61, 304, 1997.
42. Barnes, F.L., The effects of the early uterine environment on the subsequent development of embryo and fetus, *Theriogenology,* 53(2), 649, 2000.
43. Christians, E. et al., Evidence for the involvement of mouse heat shock factor 1 in the atypical expression of the HSP70.1 heat shock gene during mouse zygotic genome activation, *Mol. Cell Biol.*, 17, 778, 1997.
44. Khosla, S. et al., Culture of preimplantation mouse embryos affects fetal development and the expression of imprinted genes, *Biol. Reprod.*, 64, 918, 2001.

45. Lane, M. and Gardner, D.K., Ammonium induces aberrant blastocyst differentiation, metabolism, pH regulation, gene expression and subsequently afters fetal development in the mouse, *Biol. Reprod.*, 69, 1109, 2003.
46. Fedoriw, A.M. et al., Transgenic RNAi reveals essential function for CTCF in H19 gene imprinting, *Science*, 303, 238, 2004.

8 Current Concepts in Cat Cloning

Martha C. Gómez and C. Earle Pope

CONTENTS

8.1 INTRODUCTION

During the last decade, substantial progress has been made in the field of assisted reproductive techniques in domestic and nondomestic cats. Basic knowledge of embryo physiology has been accrued and reliable protocols for inducing oocyte collection, *in vitro* fertilization, *in vitro* embryo culture, and embryo transfers have been established in the domestic cat and some species of endangered felids [1–4]. Recently, the feasibility of producing viable domestic and nondomestic cat offspring by nuclear transfer (NT) has also been demonstrated [5,6]. Although embryo implantation rates were relatively low, the overall level of success was comparable to that reported for other species from which NT offspring have been produced.

Interest in the possible commercial application of NT in domestic cats to reproduce desired genotypes has been gradually increasing for cat breeders wanting a specific type of animal or for pet lovers that want a copy of their beloved animal. In the future, the association of NT with genetic modification of donor cells will allow producing genetically engineered cats that, for example, will not cause allergies (allergen-free cat) and will benefit people who otherwise would not be able to have a cat pet. Genetically engineered cats also will serve as a convenient research model for studies of several human physiological abnormalities and metabolic and genetic diseases. Combining NT technology with that of genetic modification of donor cells potentially can improve the efficacy of producing genetically identical cats that may carry genes for specific disorders for improving understanding of and treatments for human diseases.

Nuclear transfer offers another approach to preserving endangered felids. With the loss on an individual from a founder population of few animals, tissue samples can be collected and stored in anticipation of using the cells for NT into enucleated oocytes of the same or different species. Therefore, the loss of genetic variation may be avoided [7] and offers the prospect of species continuation rather than extinction.

The effectiveness and efficiency of NT depends on a variety of factors, but the primary deficiency appears to be incomplete or abnormal nuclear reprogramming. Abnormal epigenetic regulators of transcription can lead to implantation failure, fetal abnormalities, and poor health of offspring [8]. This chapter reviews recent progress, problems, and applications in feline somatic cell NT. Limited information is available on domestic cat NT and most of our research has been mainly focused on interspecies NT using African wildcat (*Felis silvestris*) somatic cells as donor nuclei and domestic cat (*Felis catus*) oocytes as recipient cytoplasts; therefore, data will be presented on intra- and interspecific NT in the same section. A separate section reviews intergeneric NT using more extreme cross-genera. The process of embryo reconstruction by NT is a multistep process; some of the steps involved will be described in more detail than others.

8.2 SOURCE OF CYTOPLASTS FOR PRODUCTION OF CLONED EMBRYOS

The outcome of NT depends on several biological and technical factors, but the intrinsic quality and source of oocytes could be a key factor in determining the

TABLE 8.1
Total Number of Oocytes Matured *In Vivo* and *In Vitro* Collected for Nuclear Transfer, 2001–2003

	Metaphase II/Total Oocytes, *n* (%)	
Year	*In Vivo*[a]	*In Vitro*[b]
2001	744/1114 (66.8)	489/917 (53.3)
2002	1304/1808 (72.1)	1237/2217 (55.8)
2003	2028/2843 (72.8)	1038/2049 (52.0)
Total	4076/5765 (70.7)	2764/5183 (53.3)

[a] Oocytes recovered from donor domestic cats after FSH-LH-stimulation.

[b] Oocytes recovered from excised, nonstimulated ovaries and matured *in vitro* for 21 to 24 h.

proportion of oocytes developing to the blastocyst stage [9,10]. Although oocytes matured *in vivo* and *in vitro* have been used as recipient cytoplasts for production of cloned animals in several species, the use of recipient cytoplasts matured *in vivo* may have beneficial effects on embryo and fetal development [11]. In the domestic cat, assisted reproductive technologies are well developed and both sources of oocytes are commonly used.

In the authors' laboratory, domestic female cats undergoing up to four cycles of gonadotropin treatment and oocyte recoveries produced an average of ~20 oocytes per recovery [12], and the multiple ovarian stimulations do not appear to affect oocyte quality adversely (Table 8.1). Although oocytes matured *in vivo* can be repetitively collected from individual donors, the process is significantly more expensive and labor intensive than the production of oocytes matured *in vitro*. Cat ovaries are easily accessible from veterinary clinics and are available throughout the year. Embryos derived from oocytes matured *in vitro* are capable of developing to the blastocyst stage *in vitro* and produce offspring after embryo transfer [13–15]; however, the efficiency of *in vitro* embryo production is lower than that of *in vivo* oocytes.

8.2.1 Length of *In Vitro* Maturation of Domestic Cat Oocytes for NT

Oocyte maturation is controlled by a variety of components, including cAMP, calcium, cell cycle proteins, growth factors, gonadotropins, purines, and steroids [16], but their effects on the maturation of domestic cat oocytes have not been well studied. A reasonably high percentage of fully grown cat oocytes (40 to 70%) undergo maturation *in vitro*, depending on the culture system employed. Most oocytes matured *in vitro* complete maturation to metaphase II during the first 24 h of culture [17], but additional oocytes will mature between 24 and 48 h [18,19]. Although the highest proportion of matured oocytes is reached between 42 and 45 h of culture [20], that interval also produces the highest percentage of degenerating oocytes [19].

In 2002, Skrzyszowska et al. [21] determined the effect of the *in vitro* maturation interval of recipient oocytes on *in vitro* development of domestic cat embryos produced by reconstitution with fetal fibroblasts. Table 8.2 shows results of development of cloned embryos produced from domestic cat cytoplasts after various intervals of *in vitro* oocyte maturation. They reported that a prolonged maturation period of 43 h affects *in vitro* development of reconstructed cat embryos, as indicated by lower fusion rates (57.5%), lower development of embryos to the morula stage (28.2%), and no blastocyst development. In contrast, when oocyte cytoplasts matured *in vitro* for 24 or 35 h were used, fusion rates (71.4 and 71.6%, respectively) and the percentage of cloned embryos developing to morula (56.0 and 34.0%, respectively) and blastocyst stages (8.0 and 8.5%, respectively) were significantly higher than those of embryos reconstructed with oocyte cytoplasts matured for 43 h. The authors suggested that a possible reason for the reduction in developmental potential was that 43-h oocytes were postmature, or "aged."

In another study [20], domestic cat oocytes matured *in vitro* for extended periods (42 to 45 h) showed morphological symptoms of aging, such as fragmentation of the polar body, debris in the perivitelline space, and abnormal metaphase plates. Similar abnormal morphological characteristics observed in aged postovulatory mouse oocytes included disappearance of the microfilament area around the meiotic spindle and chromatin disorganization [26].

TABLE 8.2
Development of Cat Cloned Embryos Produced from Domestic Cat Cytoplasts after Various Intervals of *In Vitro* Oocyte Maturation

Period (h)	Donor Nucleus	NT, *n*	Fused couplets, %	% Reconstructed Embryos Developed to: ≥2 cell	Morula	Blastocyst	Ref.
24	DC	35	71.4	80.0	56.0	8.0	21
35	DC	63	74.6	55.3	34.0	8.5	
43	DC	80	57.5	60.8	28.2	0	
32–34	DC	501	59.7–66.3	54.8–72.5	25.8–47.5	2.6–7.5	22
26–30	DC		30.4–66.7	80.0[a]		10.0[a]	24
26–28	DC		50.0–57.0	6.0–27.0[a]		2.0–7.0[a]	24
24	DC	213	78.0	73.0	14.0	2.3	25
24	AWC	234	85.3	84.6	46.3	23.1	

Notes: DC = Domestic cat; AWC = African wildcat. Percentage of reconstructed embryos that developed to ≥2-cell, morula or blastocyst are of the total number of fused couplet, except where not specified ([a]).

In a review of postovulatory aging, Fissore et al. [27] indicated that the abnormal morphological characteristics were the consequence of decreases in maturation-promoting factor (MPF) and mitogen-activated protein kinase (MAPK), and that aged oocytes also appear to have depressed mitochondrial function; this possibly results in the accumulation of reactive oxygen species, the release of proapoptotic products after activation, and subsequent fragmentation of the embryo. These observations suggest that aged oocytes had structural changes that may have compromised embryo development after NT. After maturing domestic cat oocytes *in vitro* for 24 to 34 h before NT, other research groups reported that blastocyst development was similar to that of the earlier report (Table 8.2) — an indication that oocytes matured for shorter periods of time are a better source of donor cytoplasts.

8.2.2 Oocyte Cytoplasmic Maturation

Even though oocyte nuclear maturation appears normal during the first 24 h of culture, inadequate cytoplasmic maturation has been observed [14,28]. It is not clear how oocytes acquire cytoplasmic maturation, but it seems that it involves the synthesis of proteins derived from nuclear and mitochondrial transcripts used by the oocyte or during early development of the embryo. In some species of mammals, all of the qualitative and quantitative changes in protein synthesis during oogenesis occur after germinal vesicle breakdown [29]. Oocytes utilized for *in vitro* maturation are collected during the germinal vesicle stage (GV); then, the protein synthesis occurring *in vitro* is dissimilar to that which takes place *in vivo* and may be inadequate. In fact, retarded metabolism in glycolysis [30] and inadequate cytoplasmic maturation of domestic cat oocytes matured *in vitro* have resulted in lower *in vitro* embryo development than that seen with oocytes matured *in vivo* [14].

It is known that MPF and MAPK factors are involved in the oocyte maturation process and in early chromatin remodeling of the donor nucleus after NT. Recently, Bogliolo et al. [28] determined the stages of meiotic maturation and the kinetics of MPF and MAPK factors of domestic cat oocytes at different intervals of *in vitro* culture. At the time of collection (0 h), ~95% of the oocytes were in the germinal vesicle (GV) stage and only ~5% were at the M-II stage. After 8 h of culture, ~60 to 70% of the oocytes had undergone germinal vesicle breakdown (GVBD) and, at 16 h, ~80 to 85% of the oocytes were in M-I or A-I phase, with only 20 to 25% remaining in the GV or GVBD stage.

It was also demonstrated that the MPF and MAPK activity increased during GVBD at 12 and 8 h, respectively, and oocytes matured *in vivo* had higher levels of MPF and MAPK (100 ± 7.48; 100 ± 6.32) compared to oocytes matured *in vitro* for 24 h (82.4 ± 9.37; 82.6 ± 10.5) or 40 h (60 ± 6.63; 66.4 ± 6.1, respectively; numbers = as arbitrary units). These results are evidence that prolonged maturation (40 h) of domestic cat oocytes decreased MPF and MAPK activity and the decrease is correlated to the aging of oocytes. Therefore, cytoplasmic oocyte maturation in domestic cats is compromised during a prolonged *in vitro* maturation interval, which in turn may affect nuclear remodeling after NT.

An important step involving acquisition of cytoplasmic maturation during *in vitro* maturation occurs in pig oocytes at the end of the folliculogenesis, between the GV-II stage and prior to GVBD [31]. The results described for the domestic cat suggest that, similar to pig oocytes, cytoplasmic maturation may start between the transition from GV to GVBD and that donor cytoplasts derived from oocytes that matured earlier may produce better embryos after NT.

8.2.3 Growth Factors Improved Cytoplasmic and Nuclear Maturation

The addition of growth factors to *in vitro* maturation medium has a positive effect on cytoplasmic and nuclear maturation of oocytes. The addition of epidermal growth factor (EGF) to oocyte maturation medium enhances fertilization frequency and blastocyst development after *in vitro* fertilization [32]. Recently, Kitiyanant et al. [22] also demonstrated that the addition of insulin growth factor-I (IGF-I) substantially improved the nuclear maturation rates of domestic cat oocytes that were fully surrounded by cumulus cells (195/254; 77.0%). These results not only corroborated the positive effect of growth factors, but also indicated that oocyte–granulosa cell junctions are necessary for the acquisition of nuclear maturation in the domestic cat, as previously demonstrated in mice [33].

It is known that mRNA and proteins are synthesized during oocyte growth and maturation [10] and that the level of mRNA is affected by conditions during *in vitro* maturation [34]. The maturation medium affects the relative abundance of specific gene transcripts [35] and the pattern of gene expression [36] of bovine oocytes matured *in vitro*. Although little is known about gene expression in domestic cat oocytes, it could be valuable to compare how the type of maturation (*in vitro* vs. *in vivo*) affects patterns of mRNA expression, and expression patterns of genes involved in meiotic competence and development of embryos derived from IVF or NT. Such information would not directly improve nuclear reprogramming; nevertheless, understanding molecular mechanisms of *in vitro* oocyte maturation and further enhancements to the maturation medium may increase the quality and developmental competence of oocytes available for use as recipient cytoplasts and, indirectly, may result in increased cloning efficiency.

8.2.4 Oocytes Matured *In Vivo* vs. Those Matured *In Vitro* as Recipient Cytoplasts

Because of the differences between oocytes matured in vivo and in vitro in their developmental competence after IVF or ICSI, a direct comparison was made between oocytes matured *in vitro* and *in vivo* as donor cytoplasts. Over a 3 year period, the authors found that a higher percentage of *in vivo* derived oocytes (4076/5765; 70.7%) completed meiotic maturation after gonadotropin treatment/oocyte retrieval,than that of oocytes matured *in vitro* for 24 h (2764/5183; 53.3%; $P < 0.05$; Table 8.1). Both sources of cytoplasts were enucleated and reconstituted with African wildcat fibroblasts (AWC) as donor nuclei. Higher fusion rates were found when matured *in vivo* oocytes were used as recipient cytoplasts (1696/1751; 96.8%), compared to matured *in vitro* oocytes (736/814; 90.4%; $P < 0.05$).

In contrast, cleavage frequency after reconstruction using *in vitro* matured oocytes (435/511, 85.1%) was higher than that after reconstructed with oocytes matured *in vivo* (944/1194, 79.1%; $P < 0.05$) [37]. Nonetheless, no significant differences in development to the blastocyst stage were observed between embryos reconstituted with cytoplasts matured *in vivo* or *in vitro* (60/219, 27.3% vs. 47/205, 23.0%, respectively) [25]. The similarity in the percentage of blastocysts observed between both types of cytoplasts should be interpreted cautiously because fewer embryos were allowed to remain in culture until day 7, as compared to the larger number evaluated for fusion and cleavage.

Even though methods for *in vitro* maturation of oocytes are not fully optimized, Shin and coworkers [5] reported the birth of the first domestic kitten after transfer of embryos produced by reconstitution of cumulus cells with enucleated oocytes matured *in vitro* for 26 to 30 h. Recently, the authors' laboratory demonstrated the *in vivo* competence of cloned embryos derived from cytoplasts matured *in vivo* by the production of full-term African wildcat kittens after embryo transfer to domestic cat recipients [6].

In summary, these data clearly indicate that both types of cytoplasts can be used for cloning domestic or nondomestic cats. However, if available, cytoplasts matured *in vivo* are a more appropriate source of cytoplasts for production of cloned cat embryos.

8.3 EFFECT OF DONOR NUCLEUS ON NUCLEAR TRANSFER

The cell cycle phase of the donor nucleus has multiple effects on reconstruction of the embryo and is therefore a major factor in the success of NT in mammals [38]. Although it is generally accepted that the donor nucleus must be at the G0 stage of the cell cycle, and the first successful clone of an adult animal was created from a sheep mammary gland cell induced to quiescence (G0) by serum starvation [39], a controversy still exists as to which stage of the cell cycle allows the most successful NT and what the effects are on embryo development. Campbell [40] reported that cells in G0 have greater developmental potential compared to cells in G1, S, or early G2. Furthermore, according to Wilmut et al. [39], cells in G0 have lower levels of RNA production; thus, chromatin configuration is more amenable to remodeling and reprogramming.

Conversely, viable cloned offspring have been produced with cycling cells [41] and with cells synchronized in the G1 phase [42] or M-phase [43–45]. To elucidate which stage of the cell cycle is more appropriate, Wells and coworkers [46] compared the nuclear cloning efficiency of transgenic and nontransgenic fetal bovine fibroblasts synchronized in G0 or G1 phase. Their results demonstrated that nontransgenic fibroblasts synchronized at the G0 phase gave higher percentages of calves at term than cells synchronized at G1 phase. In contrast, transgenic fibroblasts synchronized at G1 phase developed higher percentages of calves at term and higher postnatal survival than cells in G0. These results indicated that NT efficiency is affected by coordination between donor cell type and cell cycle stage — not only by phase of the cell cycle. These combined results confirm that donor cells at different phases of the cell cycle can produce viable offspring.

Another factor critical to the success of NT is coordination of the cell cycle between the donor nucleus and the recipient cytoplast. As reviewed by Campbell [40], three methods are used to maintain euploidy of reconstructed embryos:

- Transferring a donor nucleus in the G0/G1 phase (2C) into nonactivated oocytes at metaphase II (M-II)
- Transferring a donor nucleus in the G2/M phase (4C) into enucleated oocytes at MII to allow the expulsion of a polar body and return the reconstructed embryo to normal ploidy (2C)
- Preactivating or partially activating a recipient oocyte in M-II before transferring a donor nucleus at any stage of the cell cycle

Only one offspring has been reported by using oocytes activated before introducing a somatic cell [47]. In contrast, cloned offspring from many species have been produced with nonactivated M-II cytoplasts by introducing donor nuclei at G0/G1 or G2/M phases. The main reason that nonactivated M-II cytoplasts may be better recipients could be due to their high level of MPF activity, known to be required for nuclear remodeling and reprogramming after NT.

Although the question as to which stage of the cell cycle is the more appropriate is still unanswered, serum starvation is the most widely used method to synchronize cells in the G0/G1 phase prior to NT. The next part of the chapter will discuss the influence of the donor nucleus for producing domestic and nondomestic cat cloned embryos and methods for cell synchronization and their effect on embryo development after NT. All experiments were performed with nonactivated M-II oocytes as recipient cytoplasts because, as far as the authors are aware, preactivated cytoplasts have not been used for NT in the cat.

8.3.1 Cell Synchronization of Domestic and Nondomestic Cat Cells

Methods of arresting mammalian cells in the G0/G1 phase of the cell cycle include: serum starvation; contact inhibition; and reversible cycle inhibitors such as roscovitine, dimethyl sulfoxide (DMSO), or aphidicolin and butyrolactone. Some of these methods have been used to synchronize domestic and nondomestic cat cells.

Bochenek et al. [48] used flow cytometry to analyze the distribution of the cell cycle phases of domestic cat cumulus cells obtained from immature oocytes and fibroblast cells from skin samples that were actively cycling or serum starved. Slightly higher percentages of cumulus cells were synchronized at G0/G1 phase after 4 to 6 days of serum starvation (86.8%) as compared to cycling cells (74.0%); an increased percentage of fibroblast cells in the G0/G1 phase was observed as the starvation period was prolonged from 4 to 6 days (85.7%); to 10 to 12 days (93.3%); and to 14 to 18 days (93.1%).

In contrast, a study of the effect of cell cycle inhibitors on the distribution of AWC and domestic cat (DSH) fibroblast cells in the various phases of the cell cycle

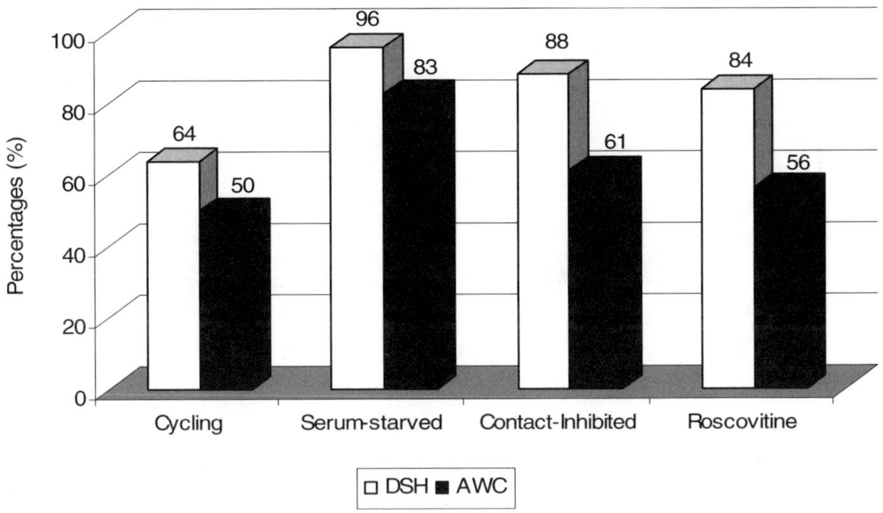

FIGURE 8.1 Percentages of domestic cat (DSH) and African wildcat (AWC) fibroblasts existing in the G0/G1 phase of the cell cycle after treatment with different cell-cycle inhibitors.

found that a higher percentage of both types of cells were synchronized after serum starvation for 5 days, compared to contact inhibition, roscovitine treatment, and actively cycling cells [25] (Figure 8.1). Similarly, when porcine and rhesus monkey cells were cultured for short periods (3 to 4 days) in serum starvation, the proportion of cells at G0/G1 phase was increased significantly [49–51].

Bochenek et al. [48] also reported that some of the domestic cat cells synchronized by the serum starvation method undergo apoptosis. Likewise, in another study, AWC and DSH fibroblast cells had higher rates of DNA fragmentation after serum starvation for 5 days. DNA damage was also observed in fetal porcine fibroblasts when the cells were cultured for a prolonged period in serum-starved medium [50]. In sheep cells, the extent of the damage was directly related to the duration of the starvation period [52].

Fetal porcine, bovine, or rhesus monkey cells cultured to 100% of confluence have a similar proportion of cells at G0/G1 as do cells after serum starvation [49,51,53], indicating that contact inhibition may be an alternative method for synchronizing cells in G0/G1; thus, the DNA damage that occurs during serum deprivation is avoided. Moreover, as compared to bovine cells synchronized by serum starvation, roscovitine produced a higher proportion of cells in G0/G1 phase and increased the nuclear reprogramming capacity and cloning efficiency of the cells, resulting in enhanced survival, both pre- and postcalving [54].

In contrast, cat fibroblast cells synchronized by contact inhibition or treated with roscovitine had a lower proportion of cells at G0/G1 phase, indicating that differences in response to cell synchronization may exist between species. In fact, the authors found that AWC fibroblast cells synchronized by different cell synchronization methods had a lower proportion of cells at the G0/G1 phase compared to DSH fibroblasts (Figure 8.1). Despite these results, it would be interesting to determine

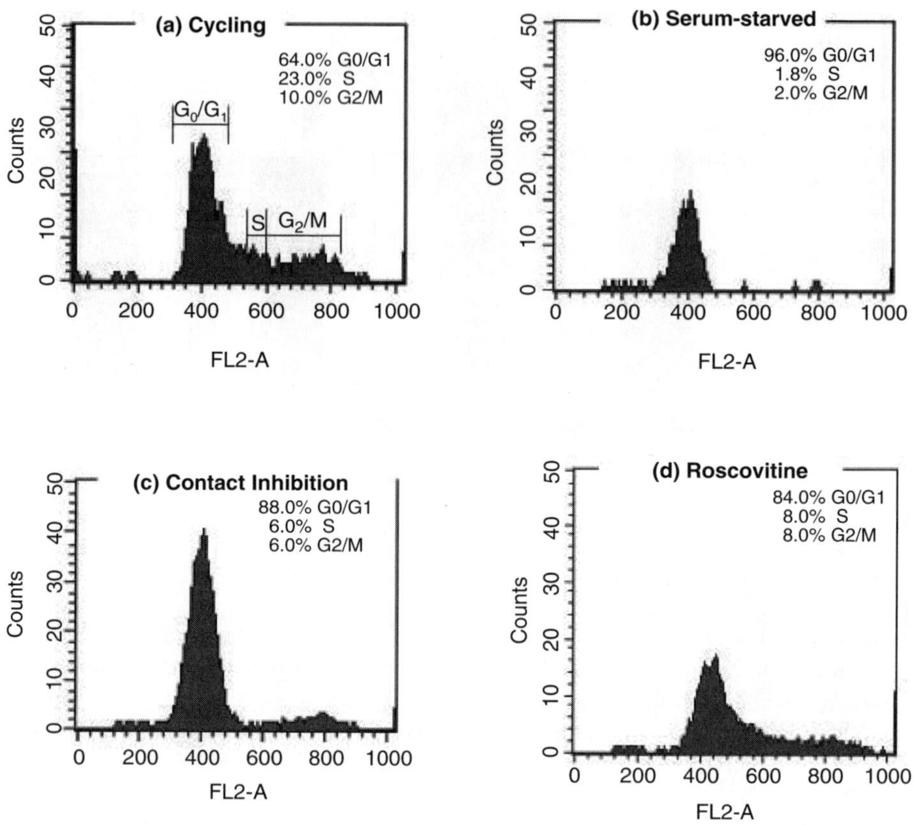

FIGURE 8.2 Typical histograms of DNA, obtained by using flow cytometry of domestic cat (DSH) fibroblast cells cultured under a variety of cell-cycle inhibitors. (From Gomez, M.C. et al., *Biol. Reprod.*, 69, 1032, 2003.)

whether roscovitine-treated cat cells would enhance kitten survival after NT, as demonstrated in the bovine.

In summary, as shown by the prominent G0/G1 DNA peaks, most domestic and nondomestic cat cells can be synchronized in the G0/G1 phase, and cycling cells have an inherently long G0/G1 phase (Figure 8.2 and Figure 8.3).

8.3.2 Effect of Cell Dissociation on the Cell Cycle and Cell Integrity

Enzymes such as trypsin are widely used to dissociate somatic cells before NT. However, exposing cells to different enzymatic methods may alter the distribution of cell cycle phases and affect cellular characteristics, such as size and DNA integrity. Different treatments for dislodging AWC and DSH fibroblast cells were evaluated to provide higher proportions of single cells and higher percentages of cells in G0/G1 phase [25]. Flow cytometry analysis revealed that enzymatic separation of AWC

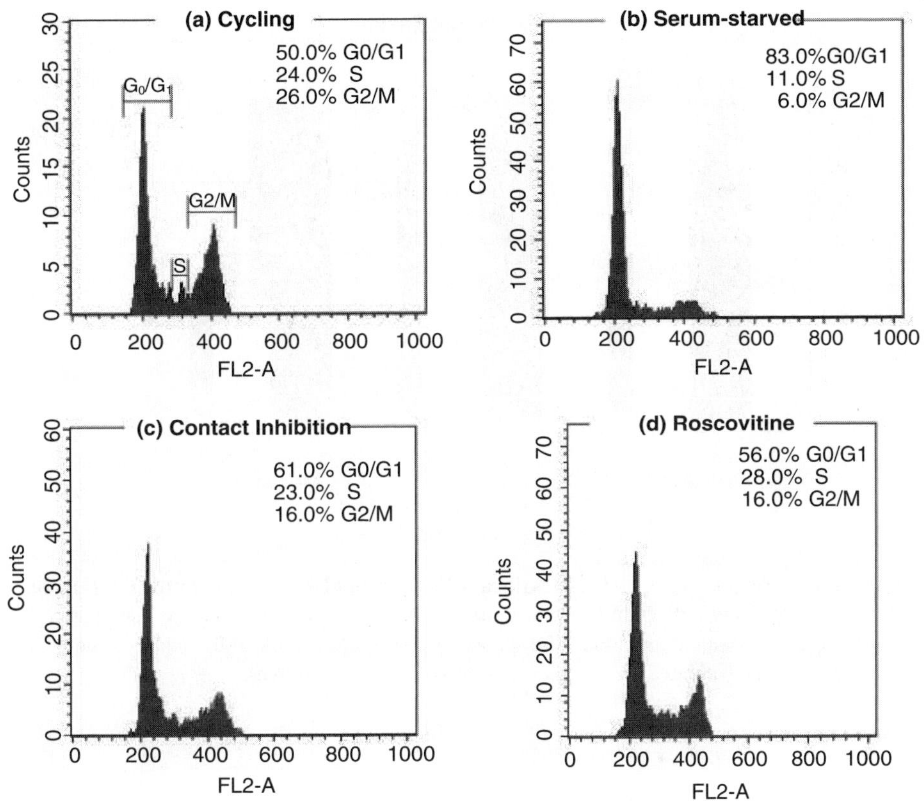

FIGURE 8.3 Typical histograms of DNA, obtained by using flow cytometry of African wildcat (AWC) fibroblast cells cultured under a variety of cell-cycle inhibitors. (From Gomez, M.C. et al., *Biol. Reprod.*, 69, 1032, 2003.)

fibroblasts with pronase or trypsin yielded higher proportions of cells in G0/G1 phase than did mechanical separation (Figure 8.4).

In contrast, mechanical separation of DSH fibroblasts yielded higher proportions of cells in G0/G1 phase compared with that from trypsin and pronase (Figure 8.5) treatment. However, in AWC and DSH cells, trypsin and pronase treatments yielded higher percentages of viable cells compared to mechanical separation (Figure 8.4 and Figure 8.5). Overall, pronase treatment yielded cells with higher viability, less cellular and DNA damage, and gave a higher percentage of cells in G0/G1 phase. Based on these results, the authors chose pronase dissociation as the preferred method for cell preparation of domestic and nondomestic cells prior to flow cytometric analysis or NT.

8.3.3 Distinguishing G0 and G1 Cell Populations

To determine whether the developmental potential of cloned embryos is improved by using donor cells in G0 phase requires the ability to distinguish G0 cells from

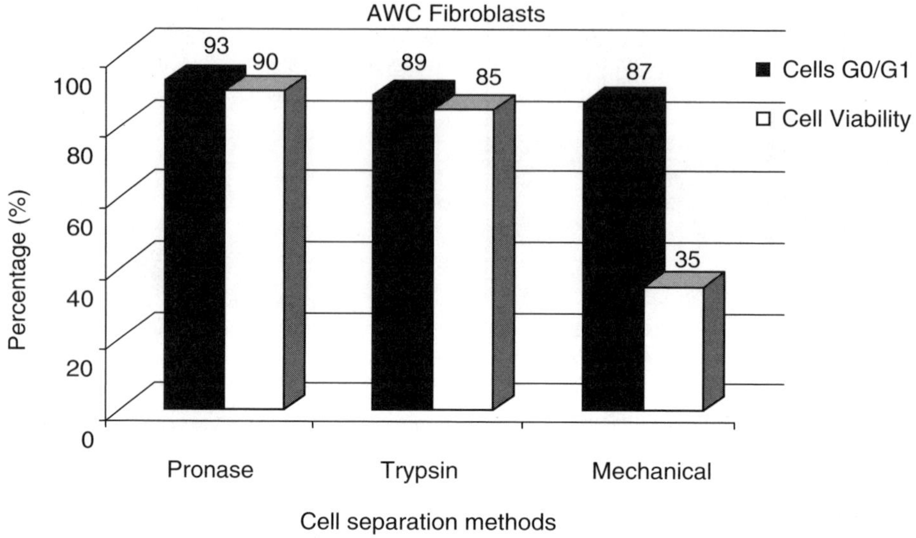

FIGURE 8.4 Percentages of African wildcat (AWC) fibroblasts existing in the G0/G1 phase of the cell cycle and cell viability after culturing cells to confluence and cell separation by enzymatic or mechanical methods. (Cell phase in G_0/G_1 was evaluated by flow cytometry and cell viability was evaluated by exclusion of eosin-B/fast green dye).

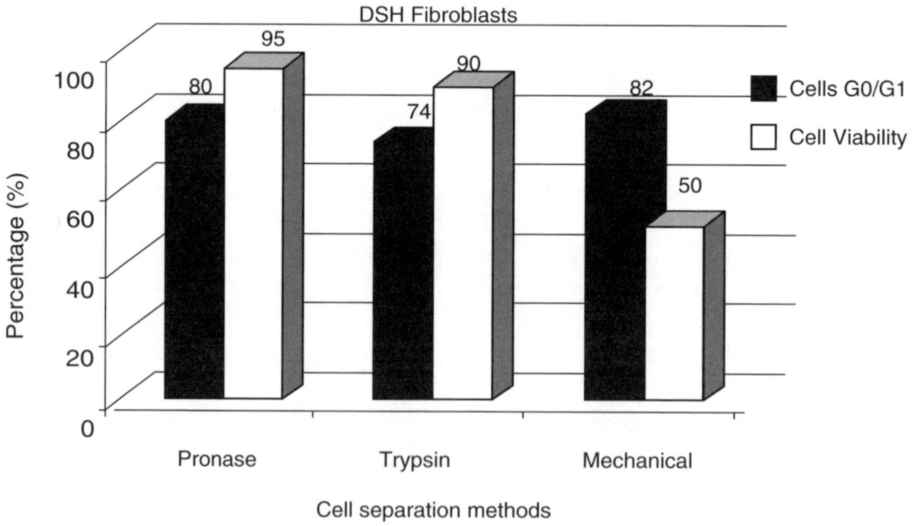

FIGURE 8.5 Percentages of domestic cat (DSH) fibroblasts existing in the G0/G1 phase of the cell cycle and cell viability after culturing cells to confluence and cell separation by enzymatic or mechanical methods. (Cell phase in G_0/G_1 was evaluated by flow cytometry and cell viability was evaluated by exclusion of eosin-B/fast green dye).

G1 cells prior to NT. Compared to G1 cells, G0 cells have much lower levels of RNA and protein [55], less expression of cyclin D [56], and lower transcript levels of interferon-τ (IF), a gene involved in early differentiation and trophoblastic function [57]. In addition to analyzing differences in these cell constituents, cell size may be a useful marker for discriminating between somatic cells in G0 and G1. For instance, higher rates of pronuclear formation occurred when small fetal fibroblasts were used for NT in pigs compared to using large fibroblasts [49].

The relationship between cell size and distribution of cells in the different phases of the cell cycle has been determined by flow cytometry in various mammalian cells after cycle synchronization using different protocols [49,58,59]. Therefore, flow cytometry can be used to measure cell size and RNA content of synchronized domestic and nondomestic cat cells to assist in discriminating cells in G0 and G1 phase. In the authors' laboratory, preliminary experiments were done to evaluate the effect of cell size on the distribution of AWC fibroblasts in the various stages of the cell cycle and the cellular RNA content of two different cell sizes [60].

Flow cytometry analyses of cell size revealed that AWC cells were distributed mainly within two subpopulations (small/medium and large). A higher percentage of serum-starved cells were classified as small/medium compared to contact-inhibited and roscovitine-treated cells (Figure 8.6). DNA content histograms of contact-inhibited and roscovitine-treated cells showed that the percentages of nuclei existing in the G0/G1 phase from the small/medium cell population were significantly higher than those for large subpopulations. However, the large cell subpopulation in the roscovitine group had a higher percentage of nuclei in the G0/G1 phase, but most of the large cells of the contact-inhibited group were in the S and G2/M phase (Figure 8.7).

To distinguish RNA content between cell populations, AWC fibroblasts were synchronized by serum starvation, contact inhibition, and roscovitine treatment, then stained with 100 μL of SYBR® Green II (Molecular Probe, SS7564). RNA content was measured using computer-generated scatterplots showing the distribution of dots with green fluorescence intensity (RNA) of individual cells in two different cell sizes. AWC fibroblasts showed a marked decline in cellular RNA levels as a consequence of serum starvation (Figure 8.8).

By comparison, scatterplots of green fluorescence of starved cells displayed a shift toward the Y axis, indicating that the majority of these cells had low RNA content. The widespread green fluorescence histogram of cells from contact-inhibited and roscovitine treated cells was more heterogeneous with respect to RNA content, indicating that both treatments had higher percentages of RNA content. When RNA content was analyzed in two separate gates (R1 = small/medium size vs. R2 = large cells), all treatments had lower percentages of RNA content in the small/medium cells than in large cells (Figure 8.8). Also, RNA content in both cell subpopulations of serum-starved cells was lower than that of contact-inhibited and roscovitine treated cells. These results clearly indicate that serum-starved cells had a lower cellular RNA content and that the small/medium cells were more likely to be in the G0 phase than were large cells.

Analysis of porcine cells by flow cytometry demonstrated that a high percentage of the small (7 to 15 μm) cell population was in G0 and G1 (88.0%) and

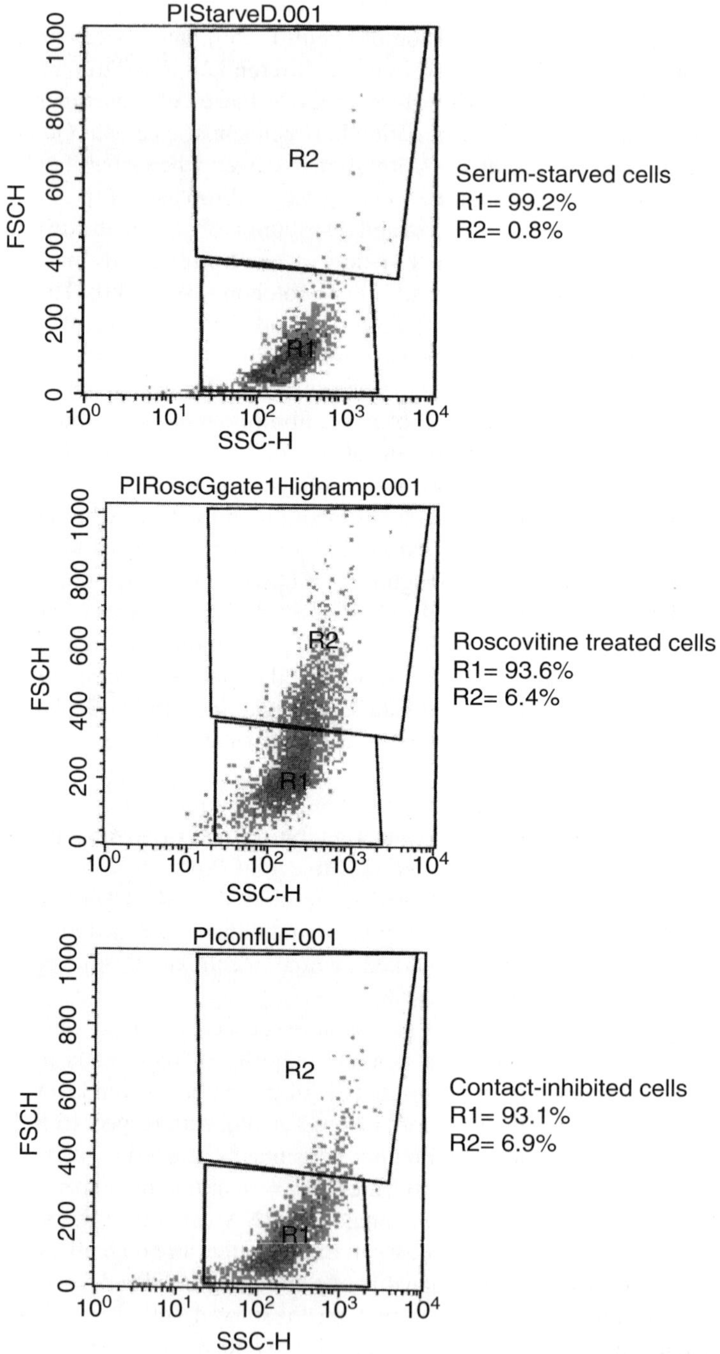

FIGURE 8.6 Distribution of AWC fibroblasts synchronized by serum-starved, contact inhibition, or treated with roscovitine, in two cell size subpopulations: R1 (small/medium) and R2 (large).

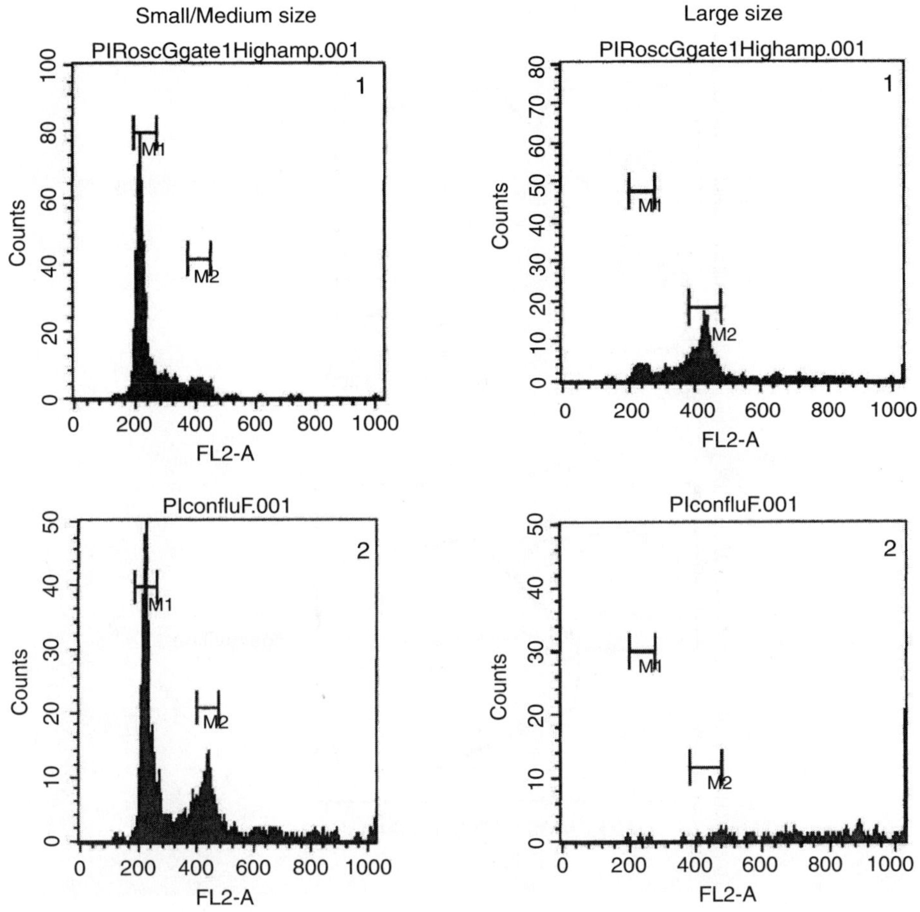

FIGURE 8.7 DNA histogram obtained by flow cytometry of African wildcat cells treated with roscovitine (1) or contact-inhibited treated (2). Cells were separated in small/medium (left column) and large (right column) cell size subpopulations. M1 indicates cells in the G0/G1 phase, and M2 indicates cells in the S and G2/M phase.

that serum-starved cells had more small cells in G0 (72.0%) than did confluent (9.0 to 16.0%) or cycling (3.0 to 7.0%) cells [49]. The authors' preliminary results were consistent with the results obtained with porcine cells in that most of the small/medium AWC cells were in the G0/G1 phase of the cell cycle. Also, serum-starved AWC cells had a higher percentage of cells in the G0 phase, as determined by their lower RNA content. Although the actual sizes of the cells in G0 and G1 were not measured in the authors' experiment or in the experiments reported by Boquest et al. [49] and Prather et al. [58], mouse T-lymphocytes cells synchronized in G0 were smaller than those in G1 [56]. The differences in cell size during the various phases of the cell cycle may be due to the decline in protein synthesis [55].

In summary, these preliminary results clearly indicate that inhibiting agents in varying degrees, influence the percentage of cells in each phase of the cell cycle,

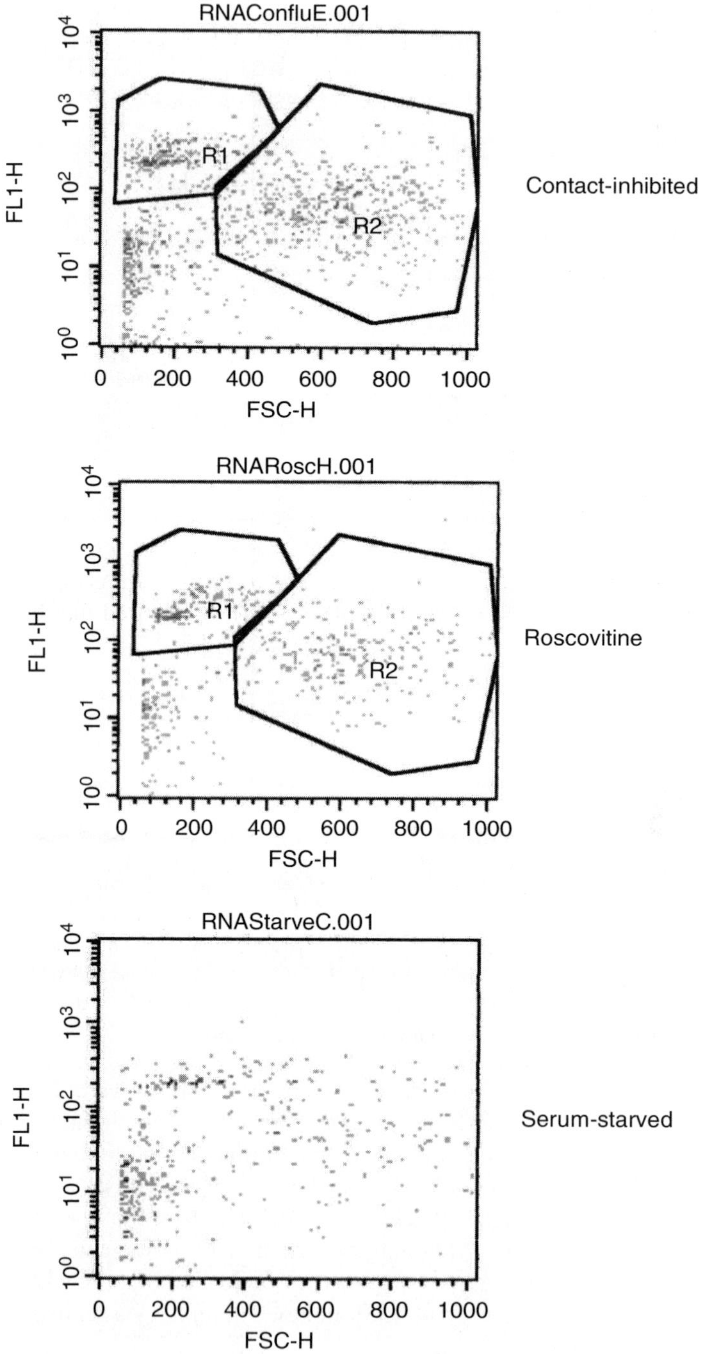

FIGURE 8.8 Flow cytometric analysis of cellular RNA content of roscovitine, contact-inhibited, and serum-starved AWC cells. Scatterplot of green fluorescence (RNA) and gating small/medium (R1) and large (R2) cells.

that the serum starvation treatment gave the highest proportion of small/medium cells, and that most of them were in the G0 and G1 phases. Therefore, additional analysis should further separate cells into two discrete sizes (small and medium) and should include counterstaining RNA vs. DNA to differentiate properly the G0 from the G1 phase.

8.3.4 Effect of Donor Nucleus Type and Cell Cycle Synchronization Method on Embryo Reconstruction and *In Vitro* Development

Table 8.3 summarizes the currently published data on embryos cloned from domestic and nondomestic cats derived after reconstruction with different type of cells that were cycling or synchronized in the G0 and G1 phases. Most research groups working in NT with the domestic cat used actively cycling cells as donor nuclei; however, the authors' routinely used serum-starved cells. Approximately 1.3 to 10.5% of cloned embryos domestic cat developed to the blastocyst stage *in vitro*, regardless of the cell type or synchronization method used.

Skrzyszowska et al. [21] evaluated the *in vitro* development of embryos cloned from domestic cats that were reconstructed using cycling cumulus cells or cycling fetal fibroblasts as the donor nucleus (Table 8.3). They reported that both types of somatic cells had similar percentages of cleavage and development to the morula or blastocyst stage, but lower fusion rates were observed when cumulus cells were

TABLE 8.3
***In Vitro* Development of Domestic and Nondomestic Cat Cloned Embryos Reconstructed with Cumulus, Muscle, Fetal, and Adult Fibroblast Cells Synchronized by Different Cell-Synchronization Methods**

Specie	Cell Type of Donor Nucleus	Donor Nucleus Synchronization Method	No. Fused/ Injected (%)	No. (%) Developing to		Ref.
				Morula	Blastocyst	
Domestic cat	Cumulus	Serum-starved 5d	82/140 (58.6)	35 (42.6)	3 (3.6)	61
Domestic cat	Fetal fibroblasts	Cycling	115/193 (59.6)	37 (32.2)	6 (5.2)	21
	Cumulus		65/143 (45.4)	20 (30.8)	5 (7.7)	
Domestic cat	Fetal fibroblasts	Cycling	80/134 (59.7)	38 (47.5)	6 (7.5)	22
	Adult fibroblasts		124/192 (64.6)	32 (25.8)	6 (4.8)	
	Cumulus		116/175 (66.3)	44 (37.9)	3 (2.6)	
Domestic cat	Muscle	Cycling	76/178 (42.7)	32 (42.1)	8 (10.5)	62
Domestic cat	Adult fibroblasts	Serum-starved 5d	75/95 (78.9)	6 (8.0)	1 (1.3)	25
		Contact inhibited	105/152 (69.1)	11 (10.5)	3 (2.8)	
		Roscovitine	95/105 (90.5)	15 (15.8)	3 (3.2)	
African wildcat	Adult fibroblasts	Serum-starved 5d	196/221 (88.7)	76 (38.8)	47 (24.0)	25
		Contact inhibited	109/129 (84.5)	45 (41.3)	22 (20.2)	
		Roscovitine	120/134 (89.5)	36 (30.0)	32 (26.7)	

used. In a similar experiment, Kitiyanant et al. [22] evaluated the development of domestic cat embryos reconstructed with cycling cumulus cells and cycling fetal or adult fibroblasts. They also found no differences in cleavage frequencies and embryo development to the morula or blastocyst stages when cumulus cells or fetal fibroblasts were used as donor nuclei. However, in contrast to the results of Skrzyszowska et al. [21], they did not find any difference in fusion rates.

Kitiyanant et al. [22] also found that cleavage rate and development of NT embryos to the morula stage were reduced after reconstruction with adult fibroblasts compared to that of NT embryos reconstructed with cumulus cells and fetal fibroblasts. Nevertheless, no differences were observed in frequency of blastocyst development of embryos reconstructed with the three types of somatic cells. The authors suggested that the possible reasons for the difference in *in vitro* developmental potential of NT embryos produced from adult fibroblast cells may due to genetic factors in the donor cells.

It has been shown that different cell types and cell synchronization methods have diverse effects on gene expression [63,64] and telomere length [65] of cloned embryos. Although substantial differences exist between differentiated adult tissues, no evidence indicates that embryo development to the blastocyst stage is improved by using a specific cell type. Some reports, however, indicate that abortions and fetal losses are higher after the transfer of bovine cloned embryos reconstructed with adult donor cells [66,67], and that epigenetic differences within cell types have a marked effect on production of cloned offspring [68]. The percentage of domestic cat blastocysts developing from each of the cell types was not significantly different, but the influence of each cell type on the percentage of normal offspring requires evaluation.

Additional work in the authors' laboratory was done to determine the influence of the method of cell synchronization on fusion, cleavage, and embryo development of domestic and nondomestic cat NT embryos. AWC and DSH cloned embryos were reconstructed with adult fibroblast cells that were synchronized in the G0/G1 phase by serum starvation, contact inhibition, and roscovitine [25]. Results indicated that the method of cell synchronization did not affect the frequency of fusion, cleavage, and blastocyst development of NT embryos. Interestingly, blastocyst development was influenced by the donor nucleus but was not affected by the cell-cycle synchronization method. In fact, higher percentages of blastocysts were observed with AWC cells (24.2%) compared to DSH cells (3.3%) (Table 8.3). These results suggest that factors other than the method of donor nuclei cell-cycle synchronization are involved in nuclear reprogramming and, in this case, the difference was due in part to the origin of the donor nucleus.

Direct comparisons of NT success between laboratories are not possible because of variations in the types of cells, methods of cell synchronization, embryo culture, and NT procedures. The percentages of blastocyst development of domestic cat embryos reconstructed with cycling fetal fibroblasts (5.2 and 7.5%), adult fibroblasts (4.8%), muscle (10.5%), and cumulus cells (5.2 and 2.6%) were slightly higher, but probably not significantly different from the rates observed with cumulus cells synchronized by serum starvation (3.6%) or adult fibroblasts synchronized by serum starvation (1.3%), contact inhibition (2.8%), or roscovitine treatment (3.2%) (Table 8.3).

Even though epigenetic differences between cell lines exist, the *in vivo* competence of domestic and nondomestic NT embryos reconstructed with cycling cumulus cells [5] and adult fibroblasts synchronized in G0/G1 phase [6] have been demonstrated.

In summary, these experiments showed that domestic and nondomestic cloned embryos can be produced by NT and that blastocyst development is not affected by cell type, although epigenetic differences between cell lines exist. In spite of the high cleavage frequency of the reconstructed embryos, blastocyst development was rather low — an indication of incomplete nuclear remodeling after NT.

8.3.5 Chromosomal Stability of Cat Somatic Cells and Derived Cloned Embryos

NT procedures are generally inefficient with less than 10% of transferred embryos surviving to term [69]. Chromosomal abnormalities, such as aneuploidy, have been associated with approximately 50% of sporadic fetal loss prior to 15 weeks in humans [70,71]. Also, in domestic animals, chromosomal imbalance appears to be an important cause of early embryonic mortality [72] and may be one of the causative factors in the higher rates of embryonic, fetal, and perinatal mortality that occur after NT. Genetic damage may occur during the *in vitro* culture of somatic cells, and a suboptimal *in vitro* culture system may contribute to an increase in the incidence of chromosomal abnormalities in somatic cells [73] and NT embryos [74].

In the domestic cat, chromosomal abnormalities were observed in FeLV-free feline tumor cell lines cultured *in vitro* [75]; unfortunately, the authors did not specify their frequency of occurrence. An analysis of the chromosomal stability of domestic cat and AWC fibroblasts cultured to different passages, and the chromosomal composition of AWC cloned blastocysts reconstructed with donor cells at different passages, was recently conducted [76,77]. To analyze the karyotypes of domestic cat fibroblasts, cells were cultured for several passages until the cells reached senescence. Karyotypic analysis showed that 16.0 to 18.0% of domestic cat fibroblasts had chromosomal abnormalities at passages 1 and 4, respectively. After the sixth passage, the percentage of aneuploid cells dramatically increased (50.1%) and by passage 9 most of the cells (87.5%) were aneuploid (Table 8.4). From the total of aneuploid cells, hypoploidy (36.0%) occurred more frequently than hyperploidy (8.6%).

Although the percentage of hyperploid cells remained stable during all passages, the proportion of hypoploid cells increased progressively with the number of passages. Similar results were observed in AWC fibroblasts after several passages. Although higher chromosomal abnormalities were observed in AWC fibroblasts at passage 1 (36.0%) and passage 4 (42.3%) compared to domestic cat cells, a higher incidence of aneuploidy was observed after the fifth passage and increased progressively until the cells reached senescence at passage 9 (97.7%) (Table 8.4). A progressive increase in chromosome aneuploidies also has been reported in porcine fibroblasts (85.0%) after prolonged culture *in vitro* [73]. However, not all somatic

TABLE 8.4
Chromosomal Abnormalities of Domestic Cat and African Wildcat Fibroblasts Cultured for Several Passages

Cell Passage No.	Aneuploid Cells Domestic Cat (%)	African Wildcat (%)
1	18.7^{a}	36.0^{a}
3	—	35.2^{a}
4	18.6^{a}	42.3^{ac}
5	33.3^{ab}	62.8^{b}
6	50.1^{ab}	56.8^{bc}
7	51.0^{ab}	77.7^{d}
8	53.9^{b}	88.8^{d}
9	87.5^{c}	97.7^{d}

Note: Different superscripts within the same column indicate significant differences (chi-square, $P < 0.05$).

cells behave similarly *in vitro*; for example, chromosomal status of sheep and bovine fibroblasts remained stable after prolonged culture *in vitro* (13.0 and 20.0%, respectively) [73,53].

Enucleated domestic cat oocytes were reconstructed with AWC fibroblasts after culture to passages 1, 3, 4, and 9 [77]. Conventional karyotyping was used to evaluate chromosomal numbers in the blastomeres. The percentage of aneuploidy in all analyzed blastomeres was similar to that of the fibroblast cells from which they were derived. When fibroblasts at passages 1, 3, and 4 with lower chromosomal abnormalities were used, similar chromosome abnormalities were found in the embryos (0, 45.0, and 31.0%, respectively). However, when donor cells at passage 9 were used, most of the blastomeres analyzed had chromosomal abnormalities (93.0%) [77]. The results indicated that the overall incidence of chromosomal abnormalities of NT embryos was correlated with the frequency at which abnormalities were observed in the donor cells.

Similar results also have been reported after analyzing bovine NT embryos with conventional karyotyping and fluorescence *in situ* hybridization (FISH) with painting probes specific for bovine X and Y chromosomes [78]. When NT bovine embryos were reconstructed with donor cells that had low chromosomal abnormalities, only 12.0% of the reconstructed embryos were abnormal. In contrast, when the embryos were reconstructed using donor cells with the highest incidence of chromosomal abnormalities (23.0%), a higher frequency of total anomalies was observed in NT embryos (55.0%).

Hyperploid embryos may be a result of cytokinetic failure, endoreplication, or blastomere fusion [79]. Conversely, most hypoploidy abnormalities are generated by the failure of sister chromatids in mitosis or paired chromosomes in meiosis to migrate to opposite poles at cell division [80], and may be a consequence of *in vitro* culture.

Accordingly, based on these results for NT, the authors currently use domestic and nondomestic cat donor cells at early passages when the percentage of cells with chromosomal abnormalities is low. It is recommended that the chromosomal stability of each cell line be analyzed prior to NT to avoid or reduce the incidence of chromosomal anomalies that may lead to lower cloning efficiency. The use of more advanced techniques, such as finding chromosome-specific probes for FISH in the cat, will provide more informative data on individual embryos than the conventional method of karyotyping.

8.4 ACTIVATION OF DOMESTIC CAT OOCYTES

Appropriate artificial activation is an essential component of the NT procedure because initiation and regulation of events during the first cell cycle are crucial for proper reprogramming of the donor nucleus and subsequent embryonic development [81]. It is well known that, during fertilization, spermatozoa entry triggers a series of intracellular calcium oscillations [82] that activate a calmodulin-dependent protein kinase II (CAMK-II). In turn, cytostatic factor (CSF) activity is down-regulated, releasing a proteolytic degradation of cyclin B; finally, MPF and MAP kinase are inactivated [83–86], leading to a resumption of meiosis and pronuclear formation. Therefore, after NT, fused couplets must be artificially activated to replace events induced by sperm entry during the normal fertilization process and to enter the first mitotic division. This part of this review considers the methods used to activate domestic cat oocytes and the effectiveness of these activation methods in reconstructed embryos after NT.

8.4.1 Activation by Exposure to Calcium Ionophores

A wide spectrum of activating stimuli, such as calcium ionophores, ethanol, or electrical pulses, are used to induce oocyte activation in several species. Calcium ionophore A23187 has been suggested as a universal activator [87] that induces intracellular calcium release [88], as well as an influx of extracellular calcium [89]. Also, ionomycin is another widely used ionophore and is believed to be more effective than A23187.

In the authors' laboratory, preliminary experiments were done to determine the adequate concentration and exposure time of the ionophores necessary to induce parthenogenetic activation of domestic cat oocytes [90]. Mature oocytes were exposed to two ionophores (A23187 or ionomycin) at two different concentrations (5 or 10 μM) for 3 or 5 min. Results indicated that the activation rate — defined as the percentage of oocytes containing one or two polar bodies and one or two pronuclei — was higher after activating oocytes matured *in vivo* (60%) than oocytes matured *in vitro* (47%). The frequency of cleavage on day 2 was higher for oocytes exposed to the calcium ionophore for 5 min than for oocytes exposed for 3 min: 42.8 vs. 23.7%, respectively ($P < 0.05$).

In contrast, cleavage frequency was not affected by the type of calcium ionophore (A23187 = 34.8% and ionomycin = 31.7%) or calcium concentration (5 μ*M* = 30.4% and 10 μ*M* = 36.1%). Blastocyst development on day 7 was not affected by any of the preceding variables, and frequencies of blastocyst development ranged from 7.1 to 16.7%. Although similar cleavage frequencies were observed after exposing oocytes to ionomicin as compared to A23187, no blastocysts were obtained when oocytes were exposed to 10 μ*M* of ionomycin for 5 min. In fact, cytogenetic analysis of oocytes matured *in vitro* that did not cleave after exposure to 10 μ*M* ionomycin for 3 or 5 min demonstrated that most oocytes exhibited abnormal chromosomal distribution, compared with oocytes exposed to the lower concentration of ionomycin.

8.4.2 Activation by Applying Electrical Stimuli

Oocyte activation has been also induced by single or multiple electrical stimuli. After applying an electrical pulse, the oocyte membrane becomes porous, facilitating entry of extracellular calcium, which, in turn, induces intracellular calcium increase and oocyte activation (see review by Macháty et al. [91]). Also, this calcium influx may possibly trigger inositol 1,4,5-triphosphate production, inducing intracellular calcium release [92].

The effect of applied voltage field strength (V/mm) and pulse number on embryo development and blastocyst formation of domestic cat oocytes matured *in vitro* was evaluated. DC pulses of 120, 140, and 160 V/mm were applied to oocytes once or twice at intervals of 0.2 sec for 60 μsec. Preliminary results indicated that the voltage field strength did not influence cleavage frequency and blastocyst development; however, the number of pulses had a direct effect on cleavage frequency [90]. A higher percentage of oocytes cleaved when they were exposed to two pulses at 120 V/mm (53%), 140 V/mm (60%), and/or 160 V/mm (80%) compared to oocytes exposed to one pulse (38, 40, and 39%, respectively; $P > 0.05$). Although cleavage frequency was higher after two pulses, blastocyst development was not different after one (2.2%) or two (2.6%) pulses.

Similar results have been reported in pigs in which oocytes were exposed to a single electrical pulse at three voltage field strengths [93]. Blastocyst development was not affected by the voltage field strength; however, when the number of pulses was increased and applied in three consecutive pulses, higher blastocyst rates were found at a low voltage field strength (1.0 kV/cc = 55.0%) than at a higher voltage field strength (1.25 kV/cm = 26.0% or 1.5 kV/cm = 35.0%).

Although preliminary results did not show an improvement in blastocyst development after applying two consecutive pulses, it is possible that the number of pulses should be increased. Several other factors that should be evaluated in cat oocytes — such as oocyte age, pulse duration, and an overall interaction among voltage field strength, pulse number, and duration [93] — influence the oocyte response to electric activation.

The preliminary results demonstrated that domestic cat oocytes can be activated by exposure to calcium ionophores or an electrical stimulus. Although either of the two calcium ionophores can be used, precaution is needed with high concentrations of ionomycin, and exposure time to either or both ionophores should not be less than 5 min. Although electric activation with two pulses enhanced cleavage rates compared to ionophores, blastocyst development was lower. It is important to mention that, although these activating stimuli can induce parthenogenetic activation, they result in reduced blastocyst development. Improvements to the electrical activation protocol may enhance the percentage of blastocysts produced.

8.4.3 Activation by Combination of Sequential Activation Treatments

The efficiency of artificial activation of oocytes can be improved by combining treatments, such as exposure to calcium activators or electrical pulses to produce an increase in intracellular calcium and a transient down regulation of MPF, with the administration of inhibitors of protein synthesis (cycloheximide, CHX) or protein phosphorylation (6-dimethylaminopurine, 6-DMAP) to induce a persistent inhibition of MPF and MPA kinase activity [94,95]. The production of cyclin B1, a component of the MPF, is prevented by CHX. The result is a drop in active MPF levels and resumption of oocyte meiosis. Inhibiting phosphorylation of a specific threonine and tyrosine, part of the MAP kinase, produces induction of oocyte activation by 6-DMAP. High levels of MAP kinase during maturation prevent activation of oocytes [96]; therefore, inhibition of protein kinases with 6-DMAP results in meiotic resumption. These combined activation protocols allow high rates of activation and support successful development of parthenogenetically activated oocytes and reconstructed embryos from several mammalian species [95,97–106].

Table 8.5 summarizes the development of domestic cat oocytes parthenogenetically activated by combined activation treatments. In the domestic cat, electrical stimuli followed by 4 to 7 h culture in 10 μg/ml CHX or 2 m*M* 6-DMAP stimulated a higher percentage of oocytes to develop to the blastocyst stage (22.6 to 27.4%) [25, 107] than electrical stimuli alone (2.2 to 2.6%) [90]. Also, combined treatments of electrical stimuli and 6-DMAP or CHX gave higher blastocyst rates than combined treatments of calcium ionophores or ethanol and protein synthesis inhibitors (see Table 8.5). In one exception, high blastocyst rates were observed by using a combined treatment of 5 μ*M* A23187 and 2.5 m*M* 6-DMAP or 10 μg/ml CHX [108]. Similar results have been observed in rabbit and pig oocytes, in which higher activation rates were found after using a combined treatment of electrical stimuli and 6-DMAP than with electrical stimuli alone [111,112]. In another study with rat oocytes, a combined treatment of electrical stimulus and 6-DMAP gave higher rates of activation and blastocyst development than combined treatment with ethanol and 6-DMAP [105].

The blastocyst cell number was reduced in domestic cat oocytes activated with a combined treatment of calcium ionophores and CHX or 6-DMAP compared to those activated with electrical stimuli and CHX (Table 8.5). However, the cell numbers of the domestic cat blastocysts were similar to those reported in embryos cloned from African wildcat [25].

It is not clear why blastocyst cell numbers were reduced in oocytes initially treated with calcium ionophore, but a high frequency of chromosomal anomalies might explain the lower cell number of these parthenogenetic embryos. It was

TABLE 8.5
Development of Domestic Cat Oocytes Parthenogenetically Activated

Oocyte Activation Treatment		No. (%) Developing to			
Ca+ Activator	Protein Inhibitor	2–6 Cells (D2–3)	Blastocyst (D7–8)	Blastocyst Cell Number ± SEM	Ref.
5 μ*M* A23187	2 m*M* 6-DMAP	70/100 (70.0)	12 (12.0)	[a]	29
10 μ*M* A23187	2 m*M* 6-DMAP	20/37 (54.0)	5 (13.5)	54.6 ± 46.4	109
	10 μg/ml CHX	19/40 (48.0)	4 (10.0)	25.5 ± 27.8	
	25 μ*M* ROS	15/29 (52.0)	5 (17.3)	23.2 ± 11.1	
5 μ*M* ionomycin	2 m*M* 6-DMAP	18/38 (47.0)	3 (7.9)	126.6 ± 11.1	
	10 μg/ml CHX	22/42 (52.0)	3 (7.1)	20.0 ± 10.0	
	25 μ*M* ROS	6/28 (21.0)	2 (7.1)	18.0 ± 2.8	
5 μ*M* A23187	2.5 m*M* 6-DMAP	142 (88.0)	[a](45.0)	[a]	108
	10 μg/ml CHX + 2.5 μg/ml CD	143 (82.0)	[a](28.0)	[a]	
7% Ethanol	10 μg/ml CHX + 1.25 μg/ml CD	56/62 (90.3)	8 (12.9)	[a]	110
Electric pulse (*n* = 2) 120 V/mm – 60 μsec	10 μg/ml CXH + 5 μg/ml CB	[a]	[a](27.4)	129.0 ± 28.0	25
Electric pulse (*n* = 2) 3.0 kV/cm – 25 μsec	10 μg/ml CHX + 5 μg/ml CB	61/84 (72.6)	19 (22.6)	154.9 ± 47.5	107
Electric pulse (*n* = 2)	2 m*M* 6-DMAP	36/43 (83.7)	11 (25.6)		90
140 V/mm – 60 μsec	10 μg/ml CHX + 5 μg/ml CB	55/65 (84.6)	12 (18.5)		

[a] Not specified.

Notes: 6-DMAP = 6-dimethylaminopurine; CHX = cycloheximide; ROS = roscovitine; CD = cytochalasin D; CB = cytochalasin B. Percentage of oocytes that developed to 2–6 cells, or blastocyst, are of the total number of oocytes activated.

previously showed that the activation of cat oocytes with a high concentration (10 μ*M*) of ionomycin induced chromosomal abnormalities. Another cause for the low cell number might be that the artificial activation protocol is unable to support repetitive and long lasting calcium oscillations. It is not clear how repetitive calcium oscillations may be beneficial for embryonic development, but, in the mouse, the persistence of calcium oscillations is closely correlated to the number of inner cell mass cells and to implantation rate [113]. Moreover, calcium oscillations may stimulate development by facilitating expression of developmentally relevant genes [114] or by promoting recruitment of maternal mRNAs and post-translational modifications of proteins that may regulate early cleavage stages [115]. Thus, the inability of artificial activation protocols to support repetitive and long lasting calcium oscillations may compromise the developmental ability of cloned embryos.

In summary, domestic cat oocytes can be partially activated by chemical or electrical stimuli. However, the combination of electrical stimulation and protein synthesis (CHX) or protein phosphorylation (6-DMAP) inhibitors is more effective and gives higher frequencies of development to the blastocyst stage.

TABLE 8.6
Development of Domestic and Nondomestic Cloned Embryos Activated by Combination of Sequential Activation Treatments

Oocyte Activation Treatment		Reconstructed Embryos Developing to		
Ca+ Activator	Protein Inhibitor	2–6 Cells (D2–3) (%)	Blastocyst (D7–8) (%)	Ref.
5 μ*M* A23187	2 m*M* 6-DMAP	54.8–72.5	2.6–7.5	29
5 μ*M* A23187	2.5 m*M* 6-DMAP	78.0	19.0	108
10 μ*M* A23187	10 μg/ml CHX	8.0–27.0	2.0–7.0	24
7% Ethanol	10 μg/ml CXH	28.2–56.0	5.2–8.0	21
7% Ethanol	10 μg/ml CHX + 1.25 μg/ml CD	85.0	11.7	110
Electric pulse (n = 2)	10 μg/ml CHX + 5 μg/ml CB	67.0–86.0	2.5–3.5	
120 V/mm — 60 μsec		79.0–89.0[a]	17.0–33.0[a]	25
Electric pulse (n = 2)	10 μg/ml CHX	80.0	10.0	23
3.0 kV/cm – 25 μsec				
Electric pulse (n = 2)	None + 10 μg/ml CB	78.9	10.5	62
1.4 kV/cm – 40 μsec				

[a] Indicates African wildcat cloned embryos.

Notes: 6-DMAP = 6-dimethylaminopurine; CHX = cycloheximide; ROS = roscovitine, CD = cytocalasin D; CB = cytocalasin B. Percent of reconstructed embryos that developed to 2–6 cells, or blastocyst, are of the total number of activated couplets.

8.4.4 Activation after Nuclear Transfer

The efficiency of a combination of sequential activation treatments on development of reconstructed embryos following NT in domestic and nondomestic cats has been demonstrated (Table 8.6). Indeed, the first domestic and African wildcat cloned kittens from somatic cells were produced by a combination of electric activation and CHX [5,6].

A comparison between activation treatments to determine possible improvements in embryo cleavage and blastocyst development after NT has not been reported in the domestic cat, but the summary data presented in Table 8.6 indicate no major influence of the activation protocol on embryo cleavage and development to the blastocyst stage. The results show that any of the combined activation protocols can be used to produce cloned embryos; however, the developmental competence of cloned embryos derived from activation protocols using calcium ionophores or ethanol has not been tested.

8.4.5 Effect of Activation on Oocyte Cytoplasmic Factors

The biochemical activity of MPF and MAP kinase of domestic cat oocytes matured *in vitro* that were chemically activated by a combination of ionomycin (5 μm for 4 min) and subsequent exposure to CHX (10 μg/ml) for 5 h was evaluated at 0, 10, and 20 h postactivation [116]. The authors reported that kinase activity of MPF and

MAP kinase was decreased at 0 h after activation and that, by 10 h, the levels of both had significantly decreased. At 20 h, the levels of MAPK had decreased even further, but the levels of MPF had risen again. Although MPF and MAP kinase activity was decreased after activation, MAP kinase activity lagged behind that of MPF.

Similar results to those reported by Keller et al. [116] were observed when kinase activity of ovine oocytes exposed to 7% ethanol was evaluated [117]. In this experiment, the decrease in MPF activity preceded the decrease of MAPK. According to the authors, a different pattern of inactivation between both kinases indicated that the MAPK inactivation was independent of MPF inactivation. In the domestic cat, the inactivation of MAP kinase was also preceded by the inactivation of MPF, and the levels of MAPK remained lower at 20 h, at which time MPF levels started to rise again. These differences in inactivation may also suggest that the inactivation of each kinase is independent of the other in the domestic cat.

The level of both kinases decreased significantly at 10 h in the domestic cat after activation compared to unactivated controls; by 20 h, concomitant with the first cleavage division, the levels of MPF had started to rise again [116]. Similar kinase activity has been observed in pig oocytes; the MPF levels in inseminated and activated oocytes remained low at 16 h and increased at 24 h, although the amplitude of decline in MPF activity was smaller for inseminated oocytes [112]. The MPF and MAP kinase activity of domestic cat oocytes fertilized *in vitro* has not been determined; however, it is possible that the kinetics of MPF and MAPK activity following combined activation treatment may be similar to that occurring following fertilization.

8.5 EMBRYO TRANSFER AND DEVELOPMENT OF CLONED EMBRYOS

To obtain the first kitten cloned from a domestic cat, 82 day-1 embryos were transferred into the oviduct of eight recipient cats (mean = 10.2 embryos per cat), from which two pregnancies (25.0%) were established. The implantation rate (2.3%) and the percentage of embryos developing to term (1.1%) were similar to those observed in other species (1.0 to 4.0%) [118]. The cloned kitten [5] was a calico female ("CC") born on December 22, 2001, after the transfer of five cloned embryos (three derived from cumulus cells and two derived from fibroblasts cells). Microsatellite analysis performed on the oral swab collected from the cloned kitten confirmed that "CC" was derived from the cumulus cell line. Despite the overall low percentage of embryos developing to term in this study, a high implantation rate was observed (20.0%) if one considers only the number of embryos ($n = 5$) transferred to the recipient that produced the term kitten. Although the other recipient cats received a mean number of 11.3 cloned embryos, they were derived from fibroblasts and the single embryo that implanted ceased development by day 44 of gestation.

As discussed earlier in Section 8.3.4, no differences in the frequency of blastocyst development of embryos reconstructed with cumulus cells or adult and fetal fibroblasts have been observed in the domestic cat (see Table 8.3). Although substantial differences exist between differentiated adult tissues, no evidence indicates that embryo development to term is improved by using a specific cell type; however epigenetic differences within cell types may have a marked effect on production of cloned offspring [68]. Few embryos were produced in the study by Shin et al. [5] and only one embryo transfer was done with NT embryos derived from cumulus cells. Nonetheless, these results may indicate that cat NT embryos derived from cumulus cells produce a higher frequency of fetal development to term compared to fibroblast cells. Clearly, the influence of cell type on the percentage of normal offspring produced after transfer of NT embryos requires evaluation.

Recently, the birth of the first wild carnivores by interspecies nuclear transfer and interspecies embryo transfer was reported [6]. The first AWC cloned kitten that survived the postnatal period was a male named Ditteaux (Figure 8.9) born on August 6, 2003. Subsequently, two additional males, known as Miles and Otis (Figure 8.10) were born on November 15, 2003. A total of 1552 AWC cloned embryos were produced by fusion of adult AWC fibroblasts with enucleated domestic cat oocytes. When fewer than 25 embryos (range = 15 to 25) were transferred into each recipient (n = 24), none of the domestic cat recipients became pregnant. However, when more than 30 embryos per recipient (range = 36.6 to 46.3) were transferred into 26 domestic cats, a total of 12 cats became pregnant (46.1%).

Of the 24 cloned embryos implanted (2.2%), nine fetuses died in uteri between days 30 and 50 of gestation and 15 embryos developed to term (>60 days; 1.5%). The embryo implantation rate and viability of AWC cloned embryos was comparable to that observed after the transfer of domestic cat embryos. Although AWC cloned embryos were derived from interspecies NT and their mitochondria may be heteroplasmic, the embryo survival rate was not reduced compared to embryo transfer results in domestic cats.

FIGURE 8.9 First African wildcat cloned kitten, "Ditteaux," with domestic cat surrogate mother.

FIGURE 8.10 African wildcat cloned kittens, "Miles" and "Otis."

Most cat embryos have been transferred by laparotomy, although a few oviductal transfers have been accomplished by laparoscopy. Whether transfers to the oviduct occurred on day 1 or to the uterus on day 5 or day 6, pregnancy rates following transfer of AWC cloned embryos were similar (50.0, 44.4, or 40.0%, respectively). However, the number of fetuses implanting after transferring embryos on day 1 (n = 17) was higher than after transferring embryos on day 5 (n = 4) or day 6 (n = 3). These results indicate that embryo transfer into the oviduct at early cleavage stages can "rescue" some NT embryos by minimizing the detrimental effects of *in vitro* culture and may increase the number of kittens produced from each transfer. Although numerous factors contribute to the low embryo viability, an incomplete reprogramming of the differentiated nucleus should be considered the primary limitation to *in vivo* development potential of cloned embryos.

Abortions are frequently observed at various stages of pregnancy after transfer of cloned embryos. For example, in cattle and sheep, one third of the pregnancies are lost during gestation [119]. The major cause of the fetal mortality [120] and an influence on neonatal defects [121] is abnormal placental development. As mentioned previously, at least one third of the AWC cloned fetuses ceased development between days 30 and 50 of gestation and it appears that placental atrophy may be the principal cause of fetal losses. In addition, placental dysfunction resulted in premature separation before delivery with a secondary consequence of respiratory failure due to lung immaturity [6].

A variety of anomalies has been reported in cloned offspring. In most cases, no comparison of the incidence of abnormalities in a comparable population is reported, and it is possible that some problems attributed to the NT process are due to other, nonrelated causes. Nevertheless, some recurring abnormalities can be found in cloned animals, such as increased birth weight, cardiovascular defects, respiratory failure, internal organ, placental, and brain defects, failure of the immune system,

TABLE 8.7
Number of Losses and Abnormal Cloned Kittens Born from Each Treatment Group

	Kittens from Early Transfer (*n*, %)	Kittens from Late Transfer (*n*, %)
Fetuses reabsorbed days 30–40	6/17 (35.3%)	1/7 (14.3%)
Died *in uteri* days 45–50 gestation	2/17 (11.7%)	0
Died *in uteri* just before delivery	2/17 (11.7%)	2/7 (28.6%)
Died by respiratory failure 1–2 h after delivery	1/17 (5.9%)	2/7 (28.6%)
Died by septicemia 36–72 h after delivery	1/17 (5.9%)	1/7 (14.3%)
Open wall cavity with organ exteriorization	3/17 (17.6%)	0

Source: Data from Gómez, M.C. et al., *Cloning Stem Cells* 6, 247, 2004.

and septicemia (see Wilmut et al. [119] for more details). Several of these disorders have also been reported in lambs and calves [122–125] produced *in vitro*.

In the authors' study, a number of abnormalities were observed in stillborn kittens and in those that died in the early postnatal period. An open wall cavity with organ exteriorization, respiratory failure, and septicemia in neonates were the main causes of death in the cloned kittens (Table 8.7). Abnormalities were found in kittens born after early (day 1) and late (day 5 or day 6) transfer; however, the open wall cavity with organ exteriorization was observed only in the early transfer group. There are not enough fetuses to determine whether differences exist between treatments. Nonetheless, live domestic and AWC cloned kittens are normal and healthy and it is important to continue evaluating these animals throughout their lives and to test their capability for reproduction.

8.6 GENE EXPRESSION AND IMPRINTING IN CLONED EMBRYOS

The efficiency of animal production by NT using somatic donor cells is still quite low. The initial success is measured by the efficiency of blastocyst development and by the production of viable offspring. However, the relatively successful development of NT embryos to the blastocyst stage *in vitro* is in contrast to their reduced competence to undergo implantation and subsequent fetal development following transfer to recipient animals [126].

For successful development of the reconstructed embryo, the donor nucleus must be reprogrammed to establish the temporal, spatial, and quantitative well-coordinated expression pattern associated with normal embryo development. Reprogramming of the nucleus constitutes a structural remodeling along with changes in genomic activity. The mechanism of reprogramming of somatic cells after NT is still unclear, but epigenetic DNA modification, such as DNA methylation [127] and changes in chromatin configuration [128] are probably involved. By definition, gene imprinting is the phenomenon in which the paternal or the maternal allele of a gene is expressed while the other allele is repressed, thus resulting in monoallelic expression of a subset of genes in the mammalian genome. Therefore, any changes in gene expression pattern can be manifested in embryo or fetal abnormalities.

To date, approximately 50 such genes have been found, with about one half expressed from the maternal allele and one half from the paternal allele [129]. The first imprinted gene to be described was insulin-like growth factor II (IGF2), whose product is a major fetal growth factor and is involved in the metabolism of early embryos. It is also a well-known survival factor that has been shown to protect a range of cell types from apoptosis [130]. In addition, it is possible that IGF2 moderates placental proliferation and that its expression is determined by the paternal:maternal allele expression ratio.

Previous studies with mouse and bovine embryos have shown that the onset of IGF2 gene transcription is markedly delayed following IVF and culture *in vitro* [131–133]. Therefore, abnormal levels of IGF2 expression may lead to failure of blastocysts to implant or to embryonic death *in utero* [134]. Overgrowth in fetuses derived from sheep embryos cultured *in vitro* before transfer to recipients has been associated with abnormal levels of IGF2 expression [135]. In humans, a genetic abnormality called Beckwith–Wiedemann syndrome (B–WS) is involved with the deregulation of several genes, including IGF2 and cyclin-dependent kinase inhibitor 1C (CDKN1C) [136]. B–WS is characterized by abdominal wall defects, visceromegaly, macroglossia, pre- and postnatal overgrowth, and neonatal hypoglycemia, and occurs at a higher incidence in babies produced by IVF [137]. Three of the authors' AWC cloned kittens were born with a major abnormality that consisted of incomplete closure of ventral body wall musculature with exteriorization of abdominal organs (Figure 8.11). Whether the body wall pathology is caused by gene deregulation or is merely a consequence of another wall anomaly is not known.

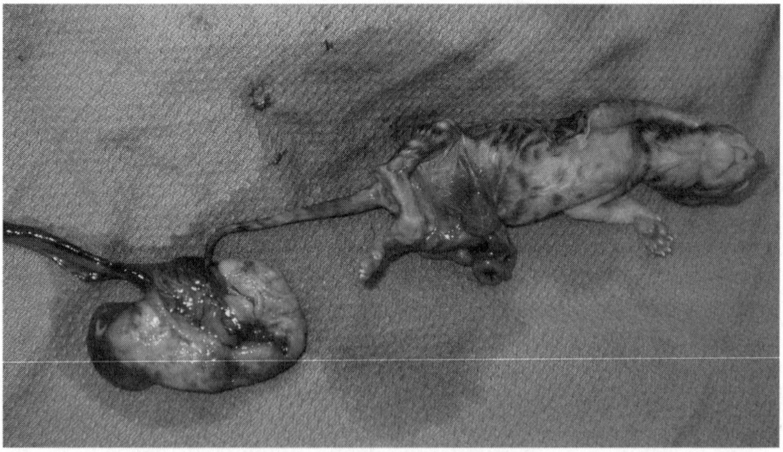

FIGURE 8.11 Abnormal African wildcat cloned with incomplete closure of ventral body wall musculature and exteriorization of abdominal organs.

8.6.1 IGF2 Gene Expression from Domestic Cat Embryos

In the domestic cat, limited information about gene expression and a lack of primer sequence information is available — particularly information about IGF2 mRNA expression. In the authors' laboratory, preliminary studies have been conducted to identify primers and to evaluate the expression of IGF2 and the housekeeping gene glyceraldehyde-3-phosphate dehydrogenase (GAPDH) [138]. A pool of ~50 domestic cat morula and/or blastocysts produced by IVF/IVC was used to isolate mRNA using a modified protocol of the PolyAT Tract System (Promega, Madison, Wisconsin). All RNA isolated was reverse transcribed and subjected to real-time PCR using primers described previously for feline GAPDH [139] and a highly conserved region of IGF2 [140].

Both sets of primers were able to amplify their target sequences from feline embryos successfully. The GAPDH primers, GAPDH.57f (5′-GCCGTFFAATTTGCCGT) and GAPDH.138r (5′-GCCATCAATGACCCCTTCAT), produced a product approximately 95 base pairs in length that was confirmed by sequencing. The IGF2 forward, IGF2f (5′GGGGACCGCGGCTTCTACTTCAG), and reverse, IGF2r (5′-GGGGTGGCACAGTACGTCTCCAG), primers successfully amplified a conserved region of the IGF2 gene (Figure 8.12). The ~130-bp

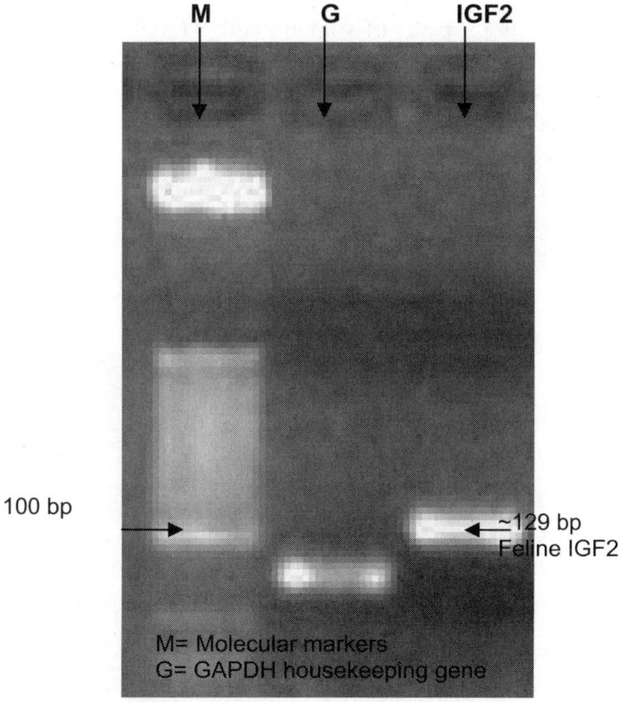

FIGURE 8.12 Analysis of IGF2, GAPDH, and molecular markers in domestic cat embryos derived from IVF.

region was confirmed by sequencing to be homologous to IGF2 in human, mouse, and pig, as well as a variety of other species.

The gene expression of GAPDH from RNA isolated from a single AWC cloned blastocyst has also been evaluated using a reverse transcription protocol adapted from Stojanov and O'Neill [132] and a low-volume RT reaction for real-time PCR utilizing the previously described primers. In addition to real time analyses, the reaction products were confirmed by electrophoresis on 3% agarose gels stained with ethidium bromide and visualized under UV light (Figure 8.13).

In summary, these preliminary results indicated that domestic cat embryos produced *in vitro* expressed IGF2 and GAPDH, and AWC cloned embryos expressed the housekeeping gene GAPDH. In further studies, it will be necessary to characterize the expression of IGF2 in AWC cloned embryos and tissues derived from abnormal cloned kittens. A detailed analysis of specific gene transcription in cloned embryos may help to determine some factors involved in the development failure. In addition, the identification of genes whose expression is frequently abnormal in cloned embryos will provide genetic markers that can be used to assess embryo viability prior to transfer to recipient animals.

8.7 INTERGENERIC NUCLEAR TRANSFER

One of the most exciting aspects of somatic cell NT is the possibility of avoiding extinction when few animals of a specific felid population remain. Intergeneric NT presents the prospect of retaining critically important genetic material by using domestic cat oocytes as recipient cytoplasts. The authors' have successfully demonstrated that interspecies NT can be achieved by using closely related species for the *Felis* genera. More extreme intergeneric combinations have not produced viable progeny to date, but have resulted in successful early embryo development *in vitro*.

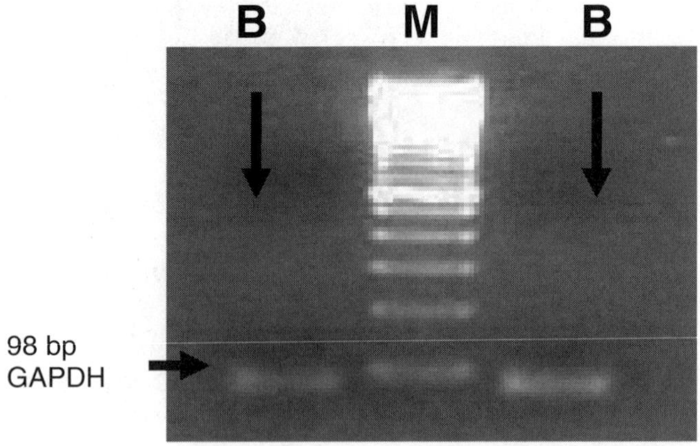

FIGURE 8.13 Analysis of GAPDH and molecular markers (M) in single AWC cloned blastocysts (B).

Fibroblast nuclei of Korean tiger (*Panthera tigris altaica*) [141]; leopard cat (*Prionailurus bengalensis*) [110]; and rusty spotted cat (*Prionailurus rubiginosus*) [142] supported blastocyst formation after transfer to enucleated domestic cat oocytes. The development of intergeneric leopard cat (6.7%) and rusty spotted cat (3.4%) cloned embryos to the blastocyst stage was lower than the development of domestic cat (13.7%) or AWC (33.3%) cloned embryos. Similarly, Korean tiger fibroblasts transplanted to bovine oocytes supported the development of embryos to the morula/blastocyst stage (1.8%) but at lower efficiency than when transplanted to domestic cat oocytes (8.8%) [141].

The low frequency of development to the blastocyst stage of these intergeneric cloned embryos may be a result of mitochondrial heteroplasmy. Indeed, embryos derived from mice that contained mitochondrial heteroplasmy had reduced ability to develop to the blastocyst stage that was associated with the foreign mitochondria [143].

Normal embryogenesis requires a correct interaction between the nucleus and the ooplasm because mitochondria (mtDNA) play an important role in supplying ATP for all energy-requiring cellular activities [144]. During normal fertilization, the sperm mitochondria are ubiquitinated and the mtDNA of the progeny is maternally inherited [145]; however, during NT, the fate of the mtDNA appears to be rather different. In interspecies cloned animals, it has been reported that the mtDNA primarily arises from the oocytes [146–148]. Thus, intergeneric macaque-rabbit [149,150] or panda-rabbit [151] cloned embryos displayed mtDNA heteroplasmy. In the case of macaque-rabbit NT embryos, mtDNA derived from the rabbit oocyte eventually became the predominant organelle at the blastocyst stage [150]. Conversely, in panda-rabbit NT embryos, mtDNA derived from panda somatic cells predominated during fetal development [151].

The mechanism regulating the distribution of foreign mtDNA in developing embryos or tissues is not known. For example, Yang and coworkers [150] demonstrated that mtDNA of macaque-rabbit cloned embryos changed their distribution during development. Both kinds of mtDNA were present at the two- to eight-cell stage, but the macaque mtDNA derived from the donor nucleus dramatically decreased after the morula stage [150]. Limited information is available about how these mtDNA changes occur and their effect on embryo and fetal viability. Nonetheless, a superior phenotype was observed when heteroplasmic cloned fetuses were produced using *Bos taurus* as donor cells and *Bos indicus* as recipient oocytes, as compared to cloned fetuses that were completely *Bos indicus* [152].

Also, mtDNA heteroplasmy was found in embryonic stem cells (ESC) obtained after fusing human somatic cells with enucleated rabbit oocytes, and the ESC maintained their heteroplasmy after several passages and were able to develop into cell types of three germ layers. More recently, healthy female mice carrying human mtDNA were able to produce normal progeny that also carried human mtDNA [153]. Therefore, interspecies mitochondrial heteroplasmy may not be an obstacle to producing viable cloned offspring.

The domestic cat has been shown to be a successful recipient of embryos from small wildcat species. Embryo transfer of nondomestic felid embryos fertilized *in vitro* has resulted in the birth of Indian desert cat and African wildcat kittens [2,154].

In addition, AWC kittens have been produced by transferring embryos derived by somatic cell NT into domestic cats [6]. Pregnancies were initiated in domestic cat recipients following transfer of fishing cat [155] fertilized *in vitro* and cloned panda embryos derived from fusing somatic nuclei of panda cells with enucleated rabbit oocytes [151]; however, development ceased during the first trimester of gestation.

Clearly, embryo implantation and fetal development after intergeneric embryo transfer can be obtained, but compromised maternal–fetal interaction is preventing fetal development to term. Although some problems associated with intergeneric NT can be overcome by improving the embryo culture system or the NT process, incompatibility between maternal–fetal interactions is more difficult to resolve. In spite of these limitations, successful intergeneric embryo transfer is possible, as demonstrated by the birth of a live healthy bongo (*Tragelaphus eurycerus*) calf after the transfer of bongo embryos into a surrogate eland recipient (*Taurotragus oryx*) [156].

8.8 CONCLUSIONS

Nuclear transfer has contributed significantly to biomedical sciences. Further development of the technology in the domestic cat will not only be important for reprogramming somatic cells from endangered species and reproducing desired genotypes, but also in dedifferentiating somatic cells that have been genetically modified.

The domestic cat is a distinctive mammalian species that exhibits a close genetic relationship to the human genome [157,158] and is a valuable research model in biomedicine for the study of several human disorders. Particularly advantageous is the possibility of combining transgenic and cloning technology to produce embryos and offspring that will carry genes for specific human disorders. This is accomplished by knocking out (KO) the specific gene on a cell line growing *in vitro* and using the KO cells as donor cells for generating clone embryos. Some improvements of NT efficiency have been made during the last decade; however, it is necessary to gain a better understanding of the molecular and biochemical aspects of nuclear reprogramming for further improving the technology and reducing some of the failures of somatic cell nuclear transfer.

Lastly, for preserving endangered species, it is relevant to consider the possibility of genetic incompatibility between the donor nucleus and the cytoplasm mitochondria in reconstructed cloned embryos and its potential effect on the successful generation of live offspring.

REFERENCES

1. Pope, C.E., Embryo technology in conservation efforts for endangered felids, *Theriogenology,* 53, 163, 2000.
2. Pope, C.E. et al., Development of *in vitro* produced African wildcat (*Felis silvestris*) embryos after cryopreservation and transfer into domestic cat recipients, *Biol. Reprod.*, 62, 321, 2000.

3. Pope, C.E. et al., Oocye retieval, *in vitro* fertilization and embryo transfer in the caracal (Caracal caracal), *Theriogenology,* 55, 397, 2001.
4. Pope, C.E., Gomez, M.C., and Dresser, B.L., *In vitro* embryo production and embryo transfer in domestic and nondomestic cats, *Theriogenology,* 2005 (in press).
5. Shin, T. et al., A cat cloned by nuclear transplantation, *Nature,* 415, 859, 2002.
6. Gómez, M.C. et al., Birth of African wildcat cloned kittens born from domestic cats, *Cloning Stem Cells,* 6, 247, 2004.
7. Seidel, G.E. Jr., Cloning, transgenesis, and genetic variance in animals, *Cloning Stem Cells,* 3, 251, 2001.
8. Trounson, A., Nuclear transfer in human medicine and animal breeding, *Reprod. Fertil. Dev.*, 13, 31, 2001.
9. Rizos, D. et al., Consequences of bovine oocyte maturation, fertilization or early embryo development *in vitro* vs. *in vivo*: implications for blastocyst yield and blastocyst quality, *Mol. Reprod. Dev.*, 61, 234, 2002.
10. Lonergan, P. et al., Oocyte and embryo quality: effect of origin, culture conditions and gene expression patterns, *Reprod. Dom. Anim.*, 38, 259, 2003.
11. Wells, D.N. et al., Production of cloned lambs from an established embryonic cell line: a comparison between *in vivo* and *in vitro* matured cytoplasts, *Biol. Reprod.*, 57, 385, 1997.
12. Pope, C.E., *In vitro* fertilization and embryo transfer in felids, in *Methods in Molecular Biology*, vol. 254: *Germ Cell Protocols*, vol. 2: *Molecular Embryo Analysis, Live imaging, Transgenesis and Cloning*; Schatten, H., Ed., Humana Press Inc., Totowa, NJ, 2003, chap. 13, 242.
13. Pope, C.E. et al., *In vitro* and *in vivo* development of embryos produced by *in vitro* maturation and *in vitro* fertilization of cat oocytes, *Reprod. Fert. Suppl.*, 51, 69, 1997.
14. Gómez, M.C. et al., Births of kittens produced by intracytoplasmic sperm injection of domestic cat oocytes matured *in vitro*, *Reprod. Fertil. Dev.,* 12, 423, 2000.
15. Gómez, M.C. et al., Development of *in vitro* matured, *in vitro* fertilized domestic cat embryos following cryopreservation, culture and transfer, *Theriogenology,* 60, 239, 2003.
16. Küpker, W., Diedrich, K., and Edwards, R.G., Principles of mammalian fertilization, *Hum. Reprod.*, 13, 20, 1998.
17. Pope, C.E. et al., Blastocyst development *in vitro* following maturation, fertilization and culture of domestic cat oocytes, *Biol. Reprod.*, 50, 179, 1994.
18. Johnston, L.A., O'Brien, S.J., and Wildt, D.E., *In vitro* maturation and fertilization of domestic cat follicular oocytes, *Gam. Res.*, 24, 343, 1989.
19. Goodrowe, K.L., Hay, M., and King, A., Nuclear maturation of domestic cat ovarian oocytes *in vitro*, *Biol. Reprod.*, 45, 466, 1991.
20. Katska-Ksiazkiewicz, L. et al., Timing of nuclear maturation of nonstored and stored domestic cat oocytes, *Theriogenology,* 59, 1567, 2003.
21. Skrzyszowska, M. et al., *In vitro* developmental competence of domestic cat embryos after somatic cloning: a preliminary report, *Theriogenology,* 58, 1615, 2002.
22. Kitiyanant, Y., Saikhun, J., and Pavasuthipaisit, K., Somatic cell nuclear transfer in domestic cat oocytes treated with IGF-I for *in vitro* maturation, *Theriogenology,* 59, 1775, 2003.
23. Westhusin, M. et al., Cloning companion animals (horses, cats, and dogs), *Cloning Stem Cells,* 5, 301, 2003.
24. Fahrudin, M. et al., The effects of culture medium on *in vitro* development of domestic cat embryos reconstructed by nuclear transplantation, *Theriogenology,* 55, 268, 2001.

25. Gómez, M.C. et al., Nuclear transfer of synchronized African wildcat somatic cells into enucleated domestic cat oocytes, *Biol. Reprod.*, 69, 1032, 2003.
26. Webb, M., Howlett, S.K., and Maro B., Parthenogenesis and cytoskeletal organization in aging mouse eggs, *J. Embryol. Exp. Morphol.,* 95, 131, 1986.
27. Fissore, R.A. et al., Mechanisms underlying oocyte activation and postovulatory ageing, *Reproduction,* 124, 745, 2002.
28. Bogliolo, L. et al., M-phase promoting factor (MPF) and mitogen activated protein kinases (MAPK) activities of domestic cat oocytes matured *in vitro* and *in vivo*, *Cloning Stem Cells,* 6, 15, 2004.
29. Briggs, D., Miller, D., and Gosden, R., Molecular biology of female gametogenesis (protein synthesis and storage), in *Molecular Biology in Reproductive Medicine*, 1st ed., Fauser, B.C.J.M, Ed., CRC Press–Parthenon Publishers, Boca Raton, Florida, 1999, chap. 12, 260.
30. Spindler, R.E., Pukazhenthi, B.S., and Wildt, D.E., Oocyte metabolism predicts the development of cat embryos to blastocyst *in vitro*, *Mol. Reprod. Dev.,* 56, 163, 2000.
31. Miyoshi, K. et al., Utility of rapidly matured oocytes as recipients for production of cloned embryos from somatic cells in the pig, *Biol. Reprod.*, 67, 540, 2002.
32. Gómez, M.C. et al., Addition of epidermal growth factor (EGF) during *in vitro* maturation of cat oocytes enhances fertilization frequency and blastocyst development *in vitro*, *Theriogenology,* 55, 472, 2001.
33. Carabatsos, M.J. et al., Oocyte–granulosa cell heterologous gap junctions are required for the coordination of nuclear and cytoplasmic meiotic competence, *Dev. Biol.*, 226, 167, 2000.
34. Pocar, P. et al., Cellular and molecular mechanisms mediating the effects of polychlorinated biphenyls on oocyte developmental competence in cattle, *Mol. Reprod. Dev.,* 60, 535, 2001.
35. Watson, A.J. et al., Impact of bovine oocyte maturation media on oocyte transcript levels, blastocyst development, cell number, and apoptosis, *Biol. Reprod.*, 62, 355, 2000.
36. Rzucidlo, S.J., Gibbons, J., and Stice, S.L., Comparison by restriction fragment differential display RT-PCR of gene expression pattern in bovine oocytes matured in the presence or absence of fetal calf serum, *Mol. Reprod. Dev.,* 59, 90, 2001.
37. Gómez, M.C. et al., unpublished data, 2004.
38. Campbell, K.H. et al., Cell cycle coordination in embryo cloning by nuclear transfer, *Rev. Reprod.*, 1, 40, 1996.
39. Wilmut, I. et al., Viable offspring derived from fetal and adult mammalian cells, *Nature,* 385, 810, 1997.
40. Campbell, K.H., Nuclear equivalence, nuclear transfer and the cell cycle (review), *Cloning,* 1, 3, 1999.
41. Cibelli, J.B. et al., Cloned transgenic calves produced from nonquiescent fetal fibroblasts, *Science,* 280, 1256, 1998.
42. Kasinathan, P. et al., Production of calves from G1 fibroblasts, *Nat. Biotechnol.*, 19, 1176, 2001.
43. Tani, T., Kato, T., and Tsunoda, Y., Direct exposure of chromosomes to nonactivated cytoplasm is effective for bovine somatic nucleus reprogramming, *Biol. Reprod.*, 64, 324, 2001.
44. Ono, Y. et al., Cloned mice from fetal fibroblast cells arrested at metaphase by a serial transfer, *Biol. Reprod.*, 64, 44, 2001.
45. Lai, L. et al., Production of α-knockout pigs by nuclear transfer cloning, *Science,* 295, 1089, 2002.

46. Wells, D.N. et al., Coordination between donor cell type and cell cycle stage improves nuclear cloning efficiency in cattle, *Theriogenology,* 59, 45, 2003.
47. Baguisi, A. et al., Production of goats by somatic cell nuclear transfer, *Nat. Biotech.*, 17, 456, 1999.
48. Bochenek, M. et al., Cell cycle analysis of somatic cell lines in the cloning of the domestic cat, *Theriogenology,* 55, 256, 2001.
49. Boquest, A.C., Billiy, N.D., and Prather, R.S., Flow cytometric cell cycle analysis of cultured porcine fetal fibroblast cells, *Biol. Reprod.*, 60, 1013, 1999.
50. Kues, W.A. et al., Cell cycle synchronization of porcine fetal fibroblasts: effects of serum deprivation and reversible cell cycle inhibitors, *Biol. Reprod.*, 62, 412, 2000.
51. Mitalipov, S.M. et al., Rhesus monkey embryos produced by nuclear transfer from embryonic blastomeres or somatic cells, *Biol. Reprod.*, 66, 1367, 2002.
52. Peura, T.T., Serum starvation can cause excessive DNA damage in sheep fetal fibroblasts, *Theriogenology,* 55, 285, 2001.
53. Kubota, C. et al., Six cloned calves produced from adult fibroblast cells after long-term culture, *Proc. Natl. Acad. Sci. USA,* 97, 990, 2000.
54. Gibbons, J. et al., Enhanced survivability of cloned calves from roscovitine-treated adult somatic cells, *Biol. Reprod.,* 66, 895, 2002.
55. Larsson, O. et al., Immediate effects of serum depletion on dissociation between growth in size and cell division in proliferating 3T3 cells, *J. Cell Physiol.,* 127, 267, 1986.
56. Pallante, B.A. et al., Comparison of developmental potential of G0 and G1 cells after nuclear transfer, *Theriogenology,* 57, 366, 2002.
57. Wrenzycki, C. et al., Nuclear transfer protocol affects messenger RNA expression patterns in cloned bovine blastocysts, *Biol. Reprod.*, 65, 309, 2001.
58. Prather, R.S., Boquest, A.C., and Day, B.N., Cell cycle analysis of cultured porcine mammary cells, *Cloning,* 1, 17, 1999.
59. Nour, M.H., Ikeda, K., and Takahashi, Y., Bovine nuclear transfer using cumulus cells derived from serum-starved and confluent cultures, *J. Reprod Dev.*, 4, 85, 2000
60. Gómez, M.C., Pope, C.E., and Jenkins, J., unpublished data, 2003.
61. Gómez, M.C. et al., *In vitro* development of enucleated domestic cat oocytes reconstructed by transfer of cumulus cells at different electrical currents, *Theriogenology,* 57, 415, 2002.
62. Wen, D.-C. et al., Comparison of developmental capacity for intra- and interspecies cloned cat (*Felis catus*) embryos, *Mol. Reprod. Dev.*, 66, 38, 2003.
63. Daniels, R. et al., Comparison of gene transcription in cloned bovine embryos produced by different nuclear transfer techniques, *Mol Reprod. Dev.*, 60, 281, 2001.
64. Miyoshi, K. et al., Improvements in cloning efficiencies may be possible by increasing uniformity in recipient oocytes and donor cells, *Biol. Reprod.*, 68, 1079, 2003.
65. Miyashita, N. et al., Remarkable differences in telomere lengths among cloned cattle derived from different cell lines, *Biol. Reprod.*, 66, 1649, 2002.
66. Kato, Y., Tani, T., and Tsunoda, Y., Cloning of calves from various somatic cell types of male and female adult, newborn and fetal cows, *J. Reprod. Fertil.*, 120, 231, 2000.
67. Niemann, H. et al., Gene expression patterns in bovine *in vitro* produced and nuclear transfer-derived embryos and their implications for early development, *Cloning Stem Cells,* 4, 29, 2002.
68. Lacham-Kaplan, O. et al., Developmental competence of nuclear transfer cow oocytes after direct injection of fetal fibroblast nuclei, *Cloning,* 2, 55, 2000.
69. Hill, J.R., Abnormal *in utero* development of cloned animals: implications for human cloning, *Differentiation,* 69, 174, 2002.

70. Burgoyne, P.S., Holland, K., and Stephens, R., Incidence of numerical chromosome anomalies in human pregnancy estimation from induced and spontaneous abortion data, *Hum. Reprod.*, 6, 555, 1991.
71. Byrne, J.L.B. and Ward, K., Genetic factors in recurrent abortion, *Clin. Obstet. Gynecol.,* 37, 693, 1994.
72. King, W.A., Chromosome abnormalities and pregnancy failure in domestic animals, *Adv. Vet. Sci. Comp. Med.,* 34, 229, 1990.
73. Denning, C. et al., Gene targeting in primary fetal fibroblasts from sheep and pig, *Cloning,* 4, 221, 2001.
74. Li, G.P. et al., Conditioned medium increases the polyploidy cell composition of bovine somatic cell nuclear-transferred blastocysts, *Reproduction,* 127, 221, 2004.
75. Gulino, S., Walen, K., and Wurster–Hill, D., Chromosome abnormalities in FeLV-free feline tumor cell lines, *29th Annu. Am. Cytogenet. Conf.*, 27, 1991.
76. Giraldo, A.M. et al., unpublished data, 2002.
77. Giraldo, A.M. et al., Chromosomal stability of African wildcat (*Felis silvestris libica*) somatic cells and cloned embryos, *Reprod. Fertil Dev.*, 16. 124, 2004.
78. Bureau, W.S. et al., Assessment of chromosomal abnormalities in bovine nuclear transfer embryos and in their donor cells, *Cloning Stem Cells,* 5, 123, 2003.
79. King, W.A., Linares, T., and Gustavsson, I., Cytogenesis of preimplantation embryos sired by bulls heterozygous for the 1/29 translocation, *Hereditas,* 94, 219, 1981.
80. Ono, S., Genetic implications of karyological instability of malignant somatic cells, *Physiol. Rev.,* 51, 496, 1971.
81. Fissore, R.A. et al., Initiation and organization of events during the first cell cycle in mammals: applications in cloning, *Cloning,* 1, 89, 1999.
82. Yanagimachi, R., Mammalian fertilization, in *The Physiology of Reproduction*, Knobil, E. and Neill, J., Eds., Raven Press, New York, 1988, 135.
83. Hashimoto, N. and Kishimoto, T., Regulation of meiotic metaphase by a cytoplasmic maturation promoting factor during mouse oocytes maturation, *Dev. Biol.*, 16, 242, 1988.
84. Masui, Y., The role of cytostatic factor (CSF) in the control of oocytes cell cycle — a summary of 20 years' study, *Dev. Growth Differ.*, 33, 543, 1991.
85. Lorca, T. et al., Calmodulin-dependent protein kinase II mediates inactivation of MPF and CSF upon fertilization of Xenopus eggs, *Nature,* 366, 270, 1993.
86. Johnson, J. et al., Calcium/calmodulin-dependent protein kinase II and calmodulin: regulators of the meiotic spindle in mouse eggs, *Dev. Biol.,* 204, 464, 1998.
87. Steinhardt, R.A. and Epel, D., Activation of sea-urchin eggs by a calcium ionophore, *Proc. Natl. Acad. Sci. USA,* 71, 1915, 1974.
88. Steinhardt, R.A., Epel, D., and Yanagimachi, R., Is calcium ionophore an universal activator for unfertilized eggs?, *Nature,* 252, 41, 1974.
89. Wang, W.H. et al., Activation of porcine oocytes with calcium ionophore: effects of extracellular calcium, *Mol. Reprod. Dev.,* 53, 99, 1999.
90. Gómez, M.C. et al., unpublished data, 2003.
91. Macháty, Z., Rickords, L.F., and Prather, R.S., Parthenogenetic activation of porcine oocytes after nuclear transfer, *Cloning,* 1, 101, 1999.
92. Ozil, J.P., The parthenogenetic development of rabbit oocytes after repetitive pulsatile electrical stimulation, *Development,* 109, 117, 1990.
93. Zhu, J. et al., Improvements of an electrical activation protocol for porcine oocytes, *Biol Reprod.*, 66, 635, 2002.
94. Szollosi, M.S. et al., Inhibition of protein kinases by 6-dimethylaminopurine accelerates the transition to interphase in activated mouse oocytes, *J. Cell. Sci.*, 104, 861, 1993.

95. Jílek, F. et al., Activation of pig oocytes using calcium ionophore: effect of protein synthesis inhibitor cycloheximide, *Anim. Reprod. Sci.*, 63, 101, 2000.
96. Picard, A. et al., Newly assembled cyclin B-cdc2 kinase is required to suppress DNA replication between meiosis I and meiosis II in starfish oocytes, *EMBO J.*, 15, 3590, 1996.
97. Loi, P. et al., Development of parthenogenetic and cloned ovine embryos: effect of activation protocols, *Biol. Reprod.*, 58, 1177, 1998.
98. Campbell, K.D., Reed, W.A., and White, K.L., Ability of integrins to mediate fertilization, intracellular calcium release, and parthenogenetic development in bovine oocytes, *Biol. Reprod.*, 62, 1702, 2000.
99. Martini, A.C. et al., *In vitro* parthenogenesis of mouse oocytes under several experimental conditions, *Zygote,* 8, 45, 2000.
100. Nakasaka, H. et al., Effective activation method with A23187 and puromycin to produce haploid parthenogenones from freshly ovulated mouse oocytes, *Zygote,* 8, 203, 2000.
101. Mitalipov, S.M., Nusser, K.D., and Wolf, D.P., Parthenogenetic activation of rhesus monkey oocytes and reconstructed embryos, *Biol. Reprod.,* 65, 253, 2001.
102. Xu, J. and Yang, X., Telomerase activity in early bovine embryos derived from parthenogenetic activation and nuclear transfer, *Biol. Reprod.*, 64, 770, 2001.
103. Yamano, S. et al., Fertilization failure and oocyte activation, *J. Med. Invest.*, 47, 1, 2001.
104. Nakagawa, K. et al., A combination of calcium ionophore and puromycin effectively produces human parthenogenones with one haploid pronucleus, *Zygote,* 9, 83, 2001.
105. Jiang, J-Y. et al., Parthenogenetic activation and subsequent development of rat oocytes *in vitro*, *Mol. Reprod. Dev.*, 61, 120, 2002.
106. Chesné, P. et al., Cloned rabbits produced by nuclear transfer from adult somatic cells, *Nature Biot.*, 20, 366, 2002.
107. Shin, T. et al., Developing an activation protocol for somatic cell nuclear transfer (SCNT) in the domestic cat, *Reprod. Fertil. Dev.,* 16, 272, 2004.
108. Du, F.L. et al., Parthenogenetic activation and somatic nuclear transfer in domestic cats using *in vitro* matured oocytes, *Theriogenology,* 57, 409, 2002.
109. King, A., Harris, R.F., and Gómez, M.C., Parthenogenetic activation of domestic cat oocytes, *Theriogenology,* 57, 703, 2002.
110. Lorthongpanich, C. et al., *In vitro* development of enucleated domestic cat oocytes reconstructed with skin fibroblasts of domestic and leopard cats, *Reprod. Fertil. Dev.*, 16, 149, 2004.
111. Yin, X.J. et al., Development of rabbit parthenogenetic oocytes and nuclear-transferred oocytes receiving cultured cumulus cells, *Theriogenology,* 54, 1469, 2000.
112. Grupen, C.G. et al., Effect 6-Dimethylaminopurine on electrically activated *in vitro* matured porcine oocytes, *Mol. Reprod. Dev.*, 62, 387, 2002.
113. Mikich-Bos, A., Whittingham, D.G., and Jones, K.T., Meiotic and mitotic Ca^{2+} oscillations affect cell composition in resulting blastocysts, *Dev. Biol.*, 182, 172, 1997.
114. Dolmetsch, R.E., Xu, K., and Lewis, R.S., Calcium oscillations increase the efficiency and specificity of gene expression, *Nature,* 392, 933, 1998.
115. Xu, Z., Kopf, G.S., and Schultz, R.M., Involvement of inositol 1,4,5-tris-phosphate-mediated Ca^{2+} release in early and late events of mouse egg activation, *Development,* 120, 1851, 1994.
116. Keller, D.L. et al., Feline oocyte activation: effects on levels of MPF and MAPK, *Theriogenology,* 55, 454, 2001.

117. Bogliolo, L. et al., Activity of maturation promoting factor (MPF) and mitogen-activated protein kinase (MAPK) after parthenogenetic activation of ovine oocytes, *Cloning,* 2, 185, 2000.
118. Wilmut, I. and Paterson, L., Somatic cell nuclear transfer, *Oncol. Res.*, 13, 303, 2003.
119. Wilmut, I. et al., Somatic cell nuclear transfer, *Nature,* 419, 583, 2002.
120. Hill, J.R. et al., Evidence for placental abnormality as the major cause of mortality in first-trimester somatic cell cloned bovine fetuses, *Biol. Reprod.*, 63, 1787, 2000.
121. Hill, J.R. et al., Clinical and pathological features of cloned of cloned transgenic calves and fetuses (13 case studies), *Theriogenology,* 51, 1451, 1999.
122. Schmidt, M. et al., Pregnancies, calves and calf viability after transfer of *in vitro* produced bovine embryos, *Theriogenology,* 46, 527, 1996.
123. Young, L.E., Sinclair, K.D., and Wilmut, I. Large offspring syndrome in cattle and sheep, *Rev. Reprod.*, 3, 155, 1998.
124. Ptak, G. et al., Follow-up lambing after transfer of *in vitro* produced embryos, *Theriogenology,* 53, 316, 2000.
125. Sinclair, K.D. et al., Aberrant fetal growth and development after *in vitro* culture of sheep zygotes, *J. Reprod. Fertil.*, 116, 177, 1999.
126. Gurdon, J.B. and Colman, A., The future of cloning, *Nature,* 402, 743, 1999.
127. Razin, A., et al., Changing methylation patterns during development, in *Nucleic Acid Methylation*, Clawson, G. Willis, D., Weissbach, A., and Jones P., Eds., Wiley–Liss, New York, 1990, 257.
128. Oide, T. et al., Comparative analysis of IGF2/H19 imprinted domain: identification of a highly conserved Dnase I hypersensitive region, *Genomics,* 24, 1, 1994.
129. Pennisi, E., Behind the scenes of gene expression, *Science,* 293, 1064, 2001.
130. Stojanov, T. and O'Neill, C., Ontogeny of expression of a receptor for platelet-activating factor in mouse preimplantation embryos and the effects of fertilization and culture *in vitro* on its expression, *Biol. Reprod.*, 60, 674, 1999.
131. Stojanov, T., Slechna, S., and O'Neill, C., *In-vitro* fertilization and culture of mouse embryos *in vitro* significantly retards the onset of insulin-like growth factor-II expression from the zygotic genome, *Mol. Hum. Reprod.*, 5, 116, 1999.
132. Stojanov, T. and O'Neill, C., *In vitro* fertilization causes epigenetic modifications to the onset of gene expression from the zygotic genome in mice, *Biol. Reprod,*. 64, 696, 2001.
133. Thompson, S.L. et al., Environmental effects on genomic imprinting in mammals, *Toxicol. Lett.,* 120, 143, 2001.
134. Goshen, R.B.-R.Z. et al., The role of genomic imprinting in implantation, *Fertil. Steril.*, 62, 903, 1994.
135. Young, L.E. et al., Epigenetic change in IGF2R is associated with fetal overgrowth after sheep embryo culture, *Nat. Genet.*, 27, 153, 2001.
136. Murrell, A. et al., An association between variants in the IGF2 gene and Beckwith–Wiedemann syndrome: interaction between genotype and epygenotype, *Hum. Mol. Genet.,* 13, 247, 2004.
137. Debaun, M.R., Niemitz, E.L., and Feinberg, A.P., Association of *in vitro* fertilization with Beckwith–Wiedemann syndrome and epigenetic alterations of LIT1 and H19, *Am. J. Hum. Genet.,* 72, 156, 2003.
138. Latino et al., unpublished data, 2002.
139. Daniels, R., Hall, V., and Trounson, A.O., Analysis of gene transcription in bovine nuclear transfer embryos reconstructed with granulose cell nuclei, *Biol. Reprod.*, 63, 1034, 2000.

140. Killian, J.K. et al., Monotreme IGF2 expression and ancestral origin of genomic imprinting, *J. Exp. Zool.*, 291, 205, 2001.
141. Hwang, W. et al., Interspecies somatic cell nuclear transfer for the production of endangered Korean tiger (Panthera Tigiris Altaica), *Theriogenology,* 55, 271, 2001.
142. Gómez, M.C. et al., Production of rusty spotted cat embryos by intergeneric nuclear transfer and another pregnancy in a domestic cat after transfer of African wildcat embryos produced by interspecies nuclear transfer, *Reprod. Fertil. Dev.,* 2004 (in press).
143. Nagao, Y. et al., Effect of different type of mitochondrial DNA on preimplantation embryonic development in the mouse, *J. Reprod. Dev.*, 44, 129, 1998.
144. Cummins, J.M., The role of maternal mitochondria during oogenesis, fertilization and embryogenesis, *Reprod. Biomed. Online,* 4, 176, 2002.
145. Sutovsky, P. et al., Ubiquitinated sperm mitochondria, selective proteolisis, and the regulation of mitochondrial inheritance in mammalian embryos, *Biol. Reprod.,* 63, 582, 2000.
146. Loi, P. et al., Genetic rescue of an endangered mammal by cross-species nuclear transfer using post-mortem somatic cells, *Nat. Biotechnol.,* 19, 962, 2001.
147. Meirelles, F.V., et al., Complete replacement of the mitochondrial genotype in a *Bos indicus* calf reconstructed by nuclear transfer to a *Bos taurus* oocyte, *Genetics,* 158, 351, 2001.
148. Lanza, R.P. et al., Cloning of an endangered species (*Bos gaurus*) using interspecies nuclear transfer cloning, *Cloning,* 2, 79, 2002.
149. Yang, C.X. et al., *In vitro* development and mitochondrial fate of macaca-rabbit cloned embryos, *Mol. Reprod. Dev.*, 65, 396, 2003.
150. Yang, C.X. et al., Quantitative analysis of mitochondrial DNAs in macaque embryos reprogrammed by rabbit oocytes, *Reproduction,* 127, 201, 2004.
151. Chen, D.Y. et al., Interspecies implantation and mitochondria fate of panda-rabbit cloned emrbyos, *Biol. Reprod.*, 67, 637, 2002.
152. Hiendleder, S. et al., Evidence for hybrid vigor by nuclear-cytoplasmic interaction in cultured cells from transmitochondrial cloned bovine fetuses, *Theriogenology,* 59, 257, 2003.
153. Sokolova, V.A. et al., Obtaining mice that carry human mitochondrial DNA transmitted to the progeny, *Mol. Reprod. Dev.*, 68, 299, 2004.
154. Pope, C.E. et al., *In vitro* fertilization in domestic and nondomestic cats including sequences of early nuclear events, *in vitro* development, cryopreservation and successful intra- and interspecies embryo transfer, *J. Reprod. Fertil.,* 47, 189, 1993.
155. Pope, C.E., et al., unpublished data, 2002.
156. Pope, C.E. and Loskutoff, N.M., *Zoo and Wild Animal Medicine — Current Therapy,* Fowler, M.E. and Miller, R.E., Eds., W. Saunder Company, Philadelphia, 1998, chap. 88.
157. Menotti–Raymond, M. et al., A genetic linkage map of microsatellites in the domestic cat (*Felis catus*), *Genomics,* 57, 9, 1999.
158. Murphy, W.J. et al., Extensive conservation of sex xhromosome organization between cat and human revealed by parallel radiation hybrid mapping, *Genome Res.*, 9, 1223, 1999.

9 Ovine Somatic Cell Nuclear Transfer: Retrospective Overview and Analysis of Epigenetic and Phenotypic Effects of Cloning Procedures

Robert Feil and Pasqualino Loi

CONTENTS

9.1 HISTORICAL NOTES: EMBRYONIC CLONING

The ovine species deserves indeed a place of honor in nuclear transplantation research. Lambs were the first large animals produced by nuclear transfer of embryonic cells into enucleated oocytes [1]. Approximately 10 years later, the same experimental procedure led to the production of the first lamb cloned from an adult somatic cell [2]. Two waves of interest followed from these major scientific breakthroughs. The first successful embryonic cloning in a farm animal species was met with favor among reproductive physiologists, and several private companies were set up to exploit embryo cloning for the multiplication of elite genotypes commercially [3]. However, low success rates rapidly revealed the empiricism of the technology, indicating that the passage from the bench to the farm had occurred too quickly.

This drop in enthusiasm following the initial failures was accompanied by an intense basic research effort that focused on the interaction between the recipient cytoplasm and the transferred nucleus. As soon as scientists started to look at the nuclear dynamics following nuclear transfer [4], it became clear that nuclear transplantation procedures had ignored basic cell cycle rules established in the 1970s [5]. These classical cell fusion studies had clearly demonstrated a dominant effect of the mitotic cytoplasm over other cell-cycle stages. Particularly, nuclei that were in any cell cycle stage (G1, S, G2) other than M-phase were found to be modified by the mitotic cytoplasm so that their chromosomes acquired a mitotic conformation — a process called premature chromosome condensation (PCC) [6].

PCC was consistently observed when embryonic cell nuclei were transferred into enucleated metaphase II oocytes [4]. However, PCC induction during S phase, a cell cycle stage at which the majority of embryonic cells are at any one time during the early embryonic cleavages, resulted in extensive DNA fragmentation [7]. Such DNA damage was thought to be responsible for the limited development of oocytes reconstructed with embryonic nuclei. Subsequently, extensive experimental work was aimed at determining the ideal combination between metaphase II oocytes, the classical recipient cytoplasts, and the transferred nucleus [7–11]. This joint effort between embryologists and cell cycle experts was precious at that time, in that various solutions were proposed to improve the viability of sheep embryos derived by nuclear transfer [12,13]. Pharmacological control of the kinase cascade involved in cell cycle progression, for instance, resulted in a considerable enhancement of nuclear transfer efficiency in sheep [14].

9.2 SOMATIC CELL NUCLEAR TRANSFER

Following the initial hurdles, by the late 1990s embryonic cell nuclear transfer had become a reliable procedure and clones composed of seven to nine normal individuals were regularly produced in advanced experimental farms [14]. Then, in 1997, the production of the first mammal by nuclear transfer of a somatic cell derived from an adult individual — Dolly, the sheep — was announced [2]. The enthusiasm following this remarkable achievement was stronger even than that evoked by cloning from embryonic cells. This breakthrough promised fantastic opportunities in the applied and basic sciences.

In reality, however, the only clear outcome following Dolly was a massive increase in the number of laboratories attempting somatic cell nuclear transfer. These combined efforts resulted in nothing but evidence of the low efficiency of the different cloning procedures [15]. Various problems, encountered worldwide [16], cooled the initial enthusiasm and (as had occurred 10 years earlier with embryonic cell cloning) basic studies that aimed at a better understanding of molecular mechanism underlying somatic cell nuclear transfer were initiated. However, this time, the task was going to be more complicated than with the embryonic cells.

Reprogramming embryonic cells that have just started the differentiation process is a relatively easy task. It is much more complicated to reprogram the cellular memory of a fully differentiated cell taken from an adult individual [17,18]. The logical outcome of incomplete reprogramming is poor development of embryos or fetuses derived from enucleated oocytes whose development is directed by a somatic cell and by frequent postnatal death in clones (Loi et al., submitted). Although the sheep was the first mammal to be cloned from adult cells, all the preceding problems, particularly the frequent postnatal death [16], are more serious than in other species (Table 9.1). Consequently, after Dolly the sheep was put down in 2003 [19], to the authors' knowledge, there are no longer any living adult cloned sheep (Figure 9.1).

The picture changes when one considers other farm animals. Cloned cattle (Chapter 4, this book), pig (Chapter 2, this book), and probably also goats develop much better once they have overcome the critical perinatal phase; subsequently, for the most part, they grow and breed normally.

9.3 EFFORTS IN OVINE SOMATIC CELL CLONING

Retrospective analysis of somatic cell nuclear transfer efforts supports the idea that complete nuclear reprogramming occurs very rarely in cloned sheep embryos. A large-scale study by the authors' group indicates that oocytes constructed with somatic cells proceed through early embryogenesis normally, and they implant at the same rate (about 70%) as *in vitro* fertilized control embryos. However, many pregnancies were lost from the second month of gestation onwards, and the final outcome in terms of development to term was only 13% (Loi et al., submitted).

The relatively high frequency of viable development to term results from skills in managing the recipient ewes carrying clones, rather than from improved nuclear reprogramming. Particularly, the authors' have observed that recipients carrying clones derived from embryonic cells displayed a number of abnormalities in their sequential behavior preceding the delivery. Therefore, they chose to stimulate the delivery by injecting exogenous hormones. Such treatments restored a normal behavioral pattern, before and after parturition, in the recipient ewes, and significantly improved the offspring outcome [20].

The predelivery behavioral abnormalities were manifest in almost all the foster mothers; therefore all the recipient ewes were primed with a combination of cortisol and estradiol 6 days before the expected day of delivery. Although this treatment partially alleviated the problem, it did not eliminate completely the complications occurring at term. In the authors' experience, some 40% of the recipient ewes developed hydroallantoids and about the same percentage showed a premature ageing of the placenta. However, relative to both complications, offspring were delivered alive, but died at different time points after birth. Macroscopic and histological examination revealed that the cause of death was likely linked to placental abnormalities: namely, hypotrophic trophoblastic epithelium and a severe reduction in its vascularization.

Such defective extra-embryonic tissue has also been described in cloned sheep fetuses [21], and early and midterm losses could potentially be attributed to these abnormalities. It was somewhat surprising to observe the same lesions in clones that developed until birth as well. In some cases, the extent of the tissue damage was so

TABLE 9.1
Developmental Abnormalities Associated with Ovine Cloning

Donor Cells	Culture	Delivery	Weight (kg)	Life Span	Clinical Observations	Ref.
Mammary epithelial cell	Yes	Vaginal	6.6	6 years	Pulmonary adenomatosis, arthritis[a]	[2]
Granulosa	Yes	Vaginal	5.2	29 days	Eschlerichia infection, abnormal heart and kidney[b]	[52]
Granulosa	Yes	Vaginal	3.8	<1 hour	Hydronephrosis, undershot jaw	
Granulosa	Yes	Vaginal	4.2	<24 hours	Flat sternum, hydronephrosis, lung abnormalities	
Granulosa	Yes	Vaginal	3.4	<1 hour	Hydronephrosis, heart with thick walls	
Granulosa	Yes	Caesarian	6.0	>15 months	Normal	
Granulosa	Yes	Caesarian	4.0	<1 hour	Flat sternum, narrow palate, abnormal bladder	
Granulosa	Yes	Caesarian	8.2	<1 hour	Flat sternum, carpal joint flexure, cystic kidney	
Granulosa	Yes	Caesarian	5.7	<1 hour	Flat sternum, carpal joint flexure, cystic kidney	
Granulosa	Yes	Caesarian	7.2	10 days	Septic polyarthritis, hydronephrosis, elongated bladder	
Granulosa	Yes	Caesarian	7.5	<1 day	Flat sternum, skeleton abnormalities	
Granulosa	Yes		4.0	<1 day	Lung tumor, edema[c]	[53]
Granulosa	Yes		8.0	<1 day	Edema	
Granulosa	No	Vaginal	4.8	<1 day	No lung expansion, hydronephrosis[d]	[48]
Granulosa	No	Vaginal	5.3	2 days	Hydronephrosis, liver degeneration	
Granulosa	No	Caesarian	4.6	4 weeks	Septic pneumonitis,	
Granulosa	No	Caesarian	4.0	7 weeks	Dilated renal pelvis, kidney and liver degeneration	
Gran. cells mouflon	No	Vaginal (induced)	3.3	7 months	Aspecific pneumonitis	

Normal birth weight range of the breeds used in these experiments:

[a] Finn dorset; 1.2 to 5 kg
[b] Not described.
[c] Not described.
[d] Sarda; 1.9 to 7 kg.
[e] Poll dorset; 3 to 9 kg.

FIGURE 9.1 Alternative strategies to improve nuclear reprogramming aim at rendering the somatic-cell chromatin more accessible to "reprogramming" factors present in the oocytes. Such treatments have a negative effect on cell viability, however. Previously, it has been shown that donor cells treated to 55 or 75°C can be reprogrammed and can sometimes develop into viable cloned lambs. The photo shows one of these lambs with its foster mother.

massive that one could wonder how these pregnancies could have progressed to term. Although not backed up by any genetic evidence, clinical and histopathological observations in clones suggest that abnormal gene expression resulting from aberrant nuclear reprogramming of somatic cells is tolerated by the fetus proper, but has a dramatic effect on the extra-embryonic membranes.

9.4 EPIGENETIC EFFECT OF CLONING PROCEDURES

It remains unknown which mechanisms are causally involved in the phenotypic abnormalities of cloned fetuses and lambs. Because the developmental abnormalities manifest well beyond implantation, however, it appears that a delayed effect of the cloning procedures occurs. Additionally, given the stochastic nature and the diversity of the phenotypic abnormalities, it seems that multiple genes or large parts of the genome could be affected — possibly, in particular, in the extra-embryonic tissues. One of the hypotheses to account for the cloning-associated abnormalities says that these are caused by aberrant epigenetic alterations that arise during the nuclear transfer procedure. Such nongenetic changes would be somatically maintained subsequently and lead to aberrant gene expression at later developmental stages.

It has been hypothesized that epigenetic alterations could affect in particular the expression of imprinted genes — a group of mammalian genes whose expression depends on whether the gene is inherited from the mother or the father [22–25]. In the mouse, several studies have been performed to test this hypothesis. Imprinted genes were indeed found to become aberrantly expressed in embryonic and extra-embryonic tissues as a consequence of embryo culture and somatic cell nuclear transfer [23,25–30]. However, imprinted genes are not the only genes that are deregulated. Recently, a large-scale gene expression study indicated that about 4%

of all genes are deregulated in placentae and livers of neonatal cloned mice derived from ES cells and cumulus cells [27].

It would appear, nevertheless, that imprinted genes are more prone to deregulation than nonimprinted genes [27], possibly because they are epigenetically repressed on one of the two alleles already. Furthermore, several imprinted genes have been shown to play key roles in extra-embryonic and fetal development, and their deregulation therefore has an impact on pre- and postnatal growth and development [31].

An important question is whether the aberrant (imprinted) gene expression at fetal and perinatal stages is due to epigenetic alterations that arise during the cloning procedure. Most imprinted genes are clustered in the genome, and their parental-allele specific repression is regulated by so-called "imprinting-control regions" (ICRs) [31,32]. These ICRs are marked by DNA methylation on one of their two parental alleles, and any aberrant change in their allelic methylation status leads to abnormal expression [23,33].

Several mouse studies have investigated the methylation status of ICRs. Aberrant DNA methylation patterns were detected in cloned mouse conceptuses, but also in fetuses obtained following *in vitro* embryo culture [23,26–30]. It remains unclear from these studies precisely when the aberrant methylation arises; whether it is a consequence of abnormal genome reprogramming following the nuclear transfer; or whether the changes are caused by other steps of the procedure, including the *in vitro* cell and embryo culture.

As a first step to explore whether imprinted genes could be responsible, at least in part, for the cloning-associated phenotypes in sheep, the authors' and others set out to determine whether imprinting is evolutionarily conserved in sheep. By carefully comparing parthenogenetic (with two maternal genomes; 14) and naturally fertilized embryos, it was established that the insulin-like growth factor-2 (*IGF2*) and IGF2-receptor (*IGF2R*) genes are imprinted in the sheep [34–36]. Other genes whose imprinted expression is conserved in sheep include *H19* and *MEST*, and the *DLK1* and *GTL2* genes [34,37,38], indicating that imprinting is largely conserved in this ruminant species. Furthermore, the ICRs that control the *IGF2*, *IGF2R*, and *H19* genes are structurally conserved in sheep and, as in mice and humans, they carry DNA methylation on one of the parental chromosomes exclusively [39,40].

Subsequent studies on cloned lambs from three independent experiments showed that the ICR controlling imprinting at the *IGF2R* gene had lost its (allelic) DNA methylation in almost all cloned lambs (Figure 9.2). In contrast, at the ICR that controls the imprinting of the *IGF2* and *H19* genes, aberrant DNA methylation was observed in only 1 of a total of 13 cloned lambs analyzed [40]. These studies establish that, also in sheep, cloning may lead to epigenetic mutations at imprinted genes and thus affect their expression and developmental phenotype. However, whereas in mice the ICR regulating the imprinted expression of *H19* and *IGF2* is particularly prone to epigenetic deregulation, in sheep the imprinting control center at *IGF2R* is affected most frequently.

It remains to be determined whether the observed loss of imprinting had occurred already in the donor cells used; whether the nuclear transfer procedure led to incomplete and aberrant chromatin reprogramming; or whether subsequent embryo culture had affected imprinted gene regulation. Relative to the latter possibility, it

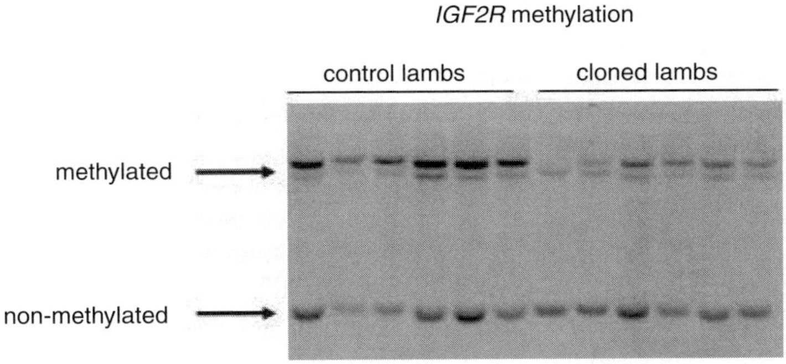

FIGURE 9.2 Deregulation of imprinting at the growth-related *IGF2R* gene in cloned lambs. *IGF2R* methylation was analyzed in tongue tissue from naturally fertilized control lambs and lambs derived by somatic cell nuclear transfer. Genomic DNA was extracted from the tissues, and was digested with the methylation-sensitive restriction enzyme *Not*I. Southern analysis was then performed using a fragment from the *IGF2R* imprinting-control region as the hybridization probe. The upper band indicates the methylated maternal allele; the lower band indicates the digested and unmethylated paternal allele. Note that the upper band is strongly reduced in intensity in most of the cloned lambs, indicating that cloning gives rise to loss of the maternal methylation imprint. (For further details, see Young, L. et al., *Mech. Dev.*, 120, 1433, 2003, from which this figure was adapted.)

is interesting to note that, on its own, culture of preimplantation sheep embryos in medium containing serum can sometimes lead to loss of *IGF2R* imprinting as well [39]. Future studies should determine to which extent other imprinted loci are affected in cloned lambs as well, and whether, as in the mouse, imprinted genes are particularly prone to epigenetic deregulation by cloning procedures. So far, no genome-wide expression studies have been reported that address this key question.

During the developmental differentiation process, the nuclei of somatic cells acquire a specialized epigenetic organization of the genome, and such epigenetic configurations ensure that the differentiation status can be somatically maintained. When differentiated cells are used for nuclear transfer, however, their nucleus needs to be reprogrammed to obtain totipotency again. Most likely, this process involves global epigenetic changes, similar to those normally observed in fertilized eggs. In different mammalian species, for instance, following fertilization DNA methylation is removed globally from the sperm-derived genome, followed by a passive decrease in DNA methylation on the maternal genome [41]. These global reductions in DNA methylation are accompanied by loss of methylation on histone H3 in the nucleosomes and have been found to be less efficient and possibly delayed in cloned bovine embryos [41–43].

To which extent the dramatic postfertilization changes in DNA methylation are essential for normal embryonic and perinatal development remains to be determined. It has been noted that the removal of DNA methylation from the sperm genome happens only in part in cattle and, apparently, not at all in sheep [43]. To ascertain later stage effects, a recent study on cattle considered the levels of DNA methylation in fetal tissues [44]. It was found that overall DNA methylation was significantly increased in liver dissected from cloned fetuses, and this correlated with abnormal crown–rump length. Although this finding contrasts with an earlier finding of reduced DNA methylation in cloned bovine fetuses [45], it emphasizes that cloning procedures may lead to abnormal DNA methylation and could thereby affect genome integrity and expression, and thus phenotype.

It is thought that the developmental abnormalities observed in clones are a consequence of abnormal nuclear reprogramming of the somatic cell nucleus. For a differentiated cell to be able to direct the development of an enucleated oocyte, all cellular memory must be erased by the oocyte's remodeling machinery. The resulting embryonic-like nucleus must re-establish the normal developmental pattern of gene expression, leading to the formation of a normal individual.

Several strategies have been suggested to maximize the extent of nuclear reprogramming in sheep cloning. These include prolonged chromosome condensation within unactivated oocytes [46], nuclear quiescence [2], and, more recently, reverse phase cloning [47]. These suggestions are backed up by serious scientific consideration; nevertheless, although sometimes effective in improving early embryonic cleavage, they do not change the ultimate fate of the clones because no significant improvement in terms of healthy offspring production has been published so far. Many cloning laboratories are starting to envisage that an efficient reprogramming strategy might be potentially achieved only by more radical remodeling or reprogramming strategies, but the risk is that such approaches affect cell viability.

Again, experimental work carried out in sheep has proved that such approaches, although leading to reduced donor cell viability, can be pursued. It was demonstrated recently, for instance, that cells heated to 55 or 75°C in order to destabilize DNA–protein complexes, retain the potential to develop into lambs after nuclear transplantation [48]. However, in spite of the improved frequency of success during pre- and postimplantation development resulting from the authors' and other groups' [47] work, the cloned lambs produced are actually still not normal (Table 9.1).

Cloning by somatic cell nuclear transfer represents a promising research platform offering opportunities in several fields, including animal breeding; transgenic animals; safeguarding endangered animal species; and, more recently, producing animal models for testing efficiency and safety of cell transplantation therapy.

Unfortunately, a robust and repeatable protocol for nuclear reprogramming is still far away. Further studies are required to determine, at different developmental stages, how the genome is organized epigenetically. Such studies should be particularly relevant at the genome level because the enzymatic complexes involved in the postfertilization chromatin reprogramming are presumed to act genome-wide. Research in this area is particularly welcomed for understanding global epigenetic modifications during normal early development — against which somatic cell-derived cloned embryos can be compared.

In parallel, it should be relevant to devise novel strategies that improve the remodeling of somatic cell nuclei. At the moment, distinct possibilities of alternative reprogramming strategies exist. One is to treat donor cells with inhibitors of histone deacetylation [49] in order to enhance global levels of histone acetylation on the chromosomes. This might render their reprogramming more efficient. Another potential strategy for ovine cloning would involve prereprogramming of somatic cell nuclei [50,51], using more extreme approaches like proteases and/or other chromatin destabilizing agents.

Genome-wide, the reprogramming should be as efficient as possible, mimicking what happens naturally after fertilization; however, the epigenetic organization at imprinted loci must remain unchanged in order not to disrupt their expression (and thus phenotype). This presents a challenging contradiction for future cloning research: how best to achieve global reprogramming of competence without affecting the epigenetic organization of specific subsets of genes, such as imprinted genes.

ACKNOWLEDGMENTS

P. Loi acknowledges Dr. Grazyna Ptak for her skillful assistance with *in vitro* embryo production, embryo transfer, and care of recipients. The research in P. Loi's lab is supported by Firb RBNE01HPMX, Cofin 2002074357, and Cofin 2003073943; R. Feil is a recipient of ARC grant funding.

REFERENCES

1. Willadsen, S.M., Nuclear transplantation in sheep embryos, *Nature,* 320, 63, 1986.
2. Wilmut, A.E. et al., Viable offspring derived from fetal and adult mammalian cells, *Nature*, 385, 810, 1997.
3. Willadsen, S.M., Cloning of sheep and cow embryos, *Genome,* 31, 956, 1989.
4. Barnes, F.L. et al., Influence of recipient oocyte cell cycle stage on DNA synthesis, nuclear envelope breakdown, chromosome constitution, and development in nuclear transplant bovine embryos, *Mol. Reprod. Dev.*, 36, 33, 1993.
5. Rao, P.N. and Johnson, R.T., Mammalian cell fusion: studies on the regulation of DNA synthesis and mitosis, *Nature,* 225, 159, 1970.
6. Johnson, R.T. and Rao, P.N., Mammalian cell fusion: induction of premature chromosome condensation in interphase nuclei, *Nature,* 226, 717, 1970.
7. Campbell, K.H., Ritchie, W.A., and Wilmut, I., Nuclear–cytoplasmic interactions during the first cell cycle of nuclear transfer reconstructed bovine embryos: implications for deoxyribonucleic acid replication and development, *Biol Reprod.*, 49, 933, 1993.
8. Collas, P. et al., Effect of donor cell cycle stage on chromatin and spindle morphology in nuclear transplant rabbit embryos, *Biol. Reprod.*, 46, 501, 1992.
9. Collas, P., Balise, J.J., and Robl, J.M., Influence of cell cycle stage of the donor nucleus on development of nuclear transplant rabbit embryos, *Biol. Reprod.,* 46, 492, 1992.
10. Smith, L.C. and Wilmut, I., Influence of nuclear and cytoplasmic activity on the development *in vivo* of sheep embryos after nuclear transplantation, *Biol Reprod.*, 40, 1027, 1989.
11. Campbell, K.H. et al., Improved development to blastocyst of ovine nuclear transfer embryos reconstructed during the presumptive S-phase of enucleated activated oocytes, *Biol. Reprod.,* 50, 1385, 1994.
12. Campbell, K.H. et al., Cell cycle coordination in embryo cloning by nuclear transfer, *Rev. Reprod.*, 1, 40, 1996.
13. Loi, P. et al., Embryo cloning in sheep: work in progress, *Theriogenology,* 48, 1, 1997.

14. Loi, P. et al., Development of parthenogenetic and cloned ovine embryos: effect of activation protocols, *Biol. Reprod.,* 58, 1177, 1998.
15. Yanagimachi, R., Cloning: experience from the mouse and other animals, *Mol. Cell Endocrinol.*, 87, 241, 2002.
16. Young, L.E., Sinclair, K.D., and Wilmut, I., Large offspring syndrome in cattle and sheep, *Rev. Reprod,*. 3, 155, 1998.
17. Tsunoda, Y. and Kato, Y., Recent progress and problems in animal cloning, *Differentiation*, 69, 158, 2002.
18. Kikyo, N. and Wolffe, A.P., Reprogramming nuclei: insights from cloning, nuclear transfer and heterokaryons, *J. Cell Sci.*, 113, 11, 2000.
19. Williams, N., Death of Dolly marks cloning milestone, *Curr. Biol.*, 18, R209, 2003.
20. Ptak, G. et al., Improving delivery and offspring viability of *in vitro*-produced and cloned sheep embryos, *Biol. Reprod.*, 67, 1719, 2002.
21. De Sousa, P.A. et al., Evaluation of gestational deficiencies in cloned sheep fetuses and placentae, *Biol. Reprod.,* 65, 23, 2001.
22. Moore, T. and Reik, W., Genetic conflict in early development: parental imprinting in normal and abnormal growth, *Rev. Reprod.*, 1, 73, 1996.
23. Dean, W. et al., Altered imprinted gene methylation and expression in completely ES cell-derived mouse fetuses: association with aberrant phenotypes, *Development*, 125, 2273, 1998.
24. Young, L.E. and Fairburn, H.R., Improving the safety of embryo technologies: possible role of genomic imprinting, *Theriogenology*, 53, 627, 2003.
25. Khosla, S. et al., Culture of preimplantation embryos and its effect on gene expression and phenotype, *Hum. Reprod. Update,* 7, 419, 2002.
26. Doherty, A.S. et al., Differential effects of culture on imprinted *H19* expression in the preimplantation embryo, *Biol. Reprod.*, 62, 1526, 2000.
27. Humpherys, D. et al., Abnormal gene expression in cloned mice derived from embryonic stem and cumulus cell nuclei, *Proc. Natl. Acad. Sci. USA*, 99, 12889, 2002.
28. Inoue, K. et al., Cloned mice, *Science,* 295, 297, 2002.
29. Mann, M.R. et al., Disruption of imprinted gene methylation and expression in cloned preimplantation stage mouse embryos, *Biol. Reprod.*, 69, 902, 2003.
30. Ogawa, H. et al., Disruption of imprinting in cloned mouse fetuses from embryonic stem cells, *Reproduction,* 126, 549, 2003.
31. Delaval, K. and Feil, R., Regulation of mammalian genomic imprinting, *Curr. Opin. Genet. Dev.*, 14, 188, 2004.
32. Reik, W. and Walter, J. , Genomic imprinting: parental influence on the genome, *Nat. Rev. Genet.,* 2, 21, 1991.
33. Li, E., Beard, C., and Jaenisch, R., Role of DNA methylation in genomic imprinting, *Nature*, 366, 362, 1993.
34. Feil, R. et al., Genomic imprinting in ruminants: allele-specific gene expression in parthenogenetic sheep, *Mamm. Genome,* 9, 831, 1998.
35. Hagemann, L.J. et al., *In vitro* and early *in vivo* development of sheep gynogenones and putative androgenones, *Mol. Reprod. Dev.*, 50 154, 1998,.
36. McLaren, R.J. and Montgomery, G.W., Genomic imprinting of the insulin-like growth factor 2 gene in sheep, *Mamm. Genome,* 10, 588, 1999.
37. Charlier, C. et al., Human–ovine comparative sequencing of a 250-kb imprinted domain encompassing the callipyge (*clpg*) locus and identification of six imprinted transcripts: *DLK1*, *DAT, GTL2*, *PEG11*, *anti-PEG11*, and *MEG8 Genome Res.*, 11, 850, 2001.

38. Lee, R.S., Depree, K.M., and Davey, H.W., The sheep (*ovis aries*) *H19* gene: genomic structure and expression patterns, from the preimplantation embryo to adulthood, *Gene,* 301, 67, 2002.
39. Young, L.E. et al., Epigenetic change in *IGF2R* is associated with fetal overgrowth after sheep embryo culture, *Nature Genet.,* 27, 153, 2001.
40. Young, L.E. et al., Conservation of *IGF2-H19* and *IGF2R* imprinting in sheep: effects of somatic cell nuclear transfer, *Mech. Dev.,* 120, 1433, 2003.
41. Santos, F. et al., Epigenetic marking correlates with developmental potential in cloned bovine preimplantation embryos, *Curr. Biol.*, 13, 1116, 2003.
42. Bourchis, D. et al., Delayed and incomplete reprogramming of chromosome methylation patterns in bovine cloned embryos, *Curr. Biol.*, 11, 1542, 2002.
43. Beaujean, N. et al., Nonconservation of mammalian preimplantation methylation dynamics, *Curr. Biol.,* 14, 266, 2004.
44. Hiendleder, S. et al., Tissue-specific elevated genomic cytosine methylation levels are associated with an overgrowth phenotype of bovine fetuses derived by *in vitro* techniques, *Biol. Reprod.*, 71, 217, 2004.
45. Cezar, G.G. et al., Genome-wide epigenetic alterations in cloned bovine fetuses. *Biol. Reprod.*, 68, 1009, 2003.
46. Campbell, K.H.S. et al., Sheep cloned by nuclear transfer of a cultured cell line, *Nature*, 380, 64, 1996.
47. Peura, T.T., Improved *in vitro* development rates of sheep somatic nuclear transfer embryos by using a reverse-order zona-free cloning method, *Cloning Stem Cells*, 1, 13, 2003.
48. Loi, P. et al., Nuclei of nonviable ovine somatic cells develop into lambs after nuclear transplantation, *Biol. Reprod.*, 67, 126, 2002.
49. Enright, B.P., Kubota, C., and Yang, X.C., Epigenetic characteristics and development of embryos cloned from donor cells treated by trichostatin A or 5-aza-2-deoxycytidine, *Biol. Reprod.*, 69, 896, 2003.
50. Collas, P., Nuclear reprogramming in cell-free extracts, *Philos. Trans. R. Soc. Lond. B Biol. Sci.,* 358, 1389, 2003.
51. Sullivan, E.J. et al., Cloned calves from chromatin remodeled *in vitro, Biol. Reprod.*, 70, 146, 2004.
52. Peura, T.T. et al., Effect of nutrition of oocyte donor on the outcomes of somatic cell nuclear transfer in the sheep, *Biol. Reprod.*, 68, 45, 2003.
53. Gou, K.M. et al., Transgenic twin lambs cloned by granulosa cells, *Cloning Stem Cells*, 1, 71, 2003.

10 Cloning in the Rat

M. Hirabayashi and S. Hochi

CONTENTS

10.1 INTRODUCTION

Since 1997, births of cloned animals following nuclear transplantation (NT) with somatic cells have been reported for several mammalian species (see other chapters in this book). Two independent protocols have been applied for producing reconstructed zygotes:

- A single donor cell is fused with an enucleated oocyte and the couplet is immediately activated (cell fusion method, e.g., in sheep [1] and cattle [2]).
- The nucleus of the donor cell is microinjected into enucleated oocytes and the couplet is activated 1 to 6 h later (Honolulu method, e.g., in mouse [3] and pig [4]).

In contrast, no cloned rat offspring have been reported until recently. Zhou et al. [5] reported the first successful production of cloned rats using differentiated fetal fibroblast cells, although their data have not yet been proven to be reproducible. The purpose of this chapter is to review the literature on the unique characteristics of rat oocytes and the earlier and latest attempts at cloning rats by somatic cell NT.

10.2 USEFULNESS OF LABORATORY RATS

Rats have been used more extensively than mice in the research fields of neuroscience, pharmacology, and toxicology [6]. More than 100 rat strains with various genetic backgrounds exist, including some useful models for human diseases. For example, the SHR (spontaneously hypertensive rat) strain and the BB (rats spontaneously developing insulin-dependent diabetes mellitus) strain are well-established models for studying cardiovascular diseases and endocrinopathy. Because of the well-understood mapping of brain functions, rats are often used for physiological studies on memory and emotion. Furthermore, experimental studies on mammary tumors require the use of rats in which symptoms of the disease are distinct from those caused by mouse mammary tumor viruses; the latter cause tumors in the mammary glands of mice but not those in humans or rats.

In addition, transgenic rats have been used as model animals for human diseases (e.g., Alzheimer's disease, autoimmunity, and HDL metabolism) and organ transplantation, and as animal bioreactors for protein production. One advantage of using rats rather than mice in transgenic studies is the ease of continuous or repeated sample collection (e.g., of blood or urine) and surgery, due to their larger size; litter size, gestation length, maturation rate, estrous cycle length, and life span of rats are very similar to those of mice. Thus, the rat has the advantage of being a reasonably well-characterized and intermediate-sized rodent that can be maintained much more cheaply than larger animals and can often be manipulated much more easily than smaller rodents.

10.3 CHARACTERISTICS OF RAT OOCYTES

10.3.1 Spontaneous Activation, Unique to This Species

Ovulated rat oocytes have been reported to activate spontaneously during *in vitro* culture [7]. Briefly, the activated oocytes extruded the second polar body within 60 to 90 min of culture; however, development became arrested again at the stage known as metaphase-III without formation of a pronucleus (PN). When reaching the metaphase-III stage, oocytes exhibited a very low rate of first cleavage following chemically induced activation. The factors involved in the spontaneous activation of rat oocytes are the amount of time the oocytes are left in the oviducts before recovery from the donor rats and the temperature used for oocyte manipulation following recovery. Reducing the time of oocyte recovery to less than 5 min after sacrificing the donor rats and handling of oocytes at 37°C reduced the spontaneous activation of the rat oocytes.

The authors have confirmed and expanded these findings; most of the denuded oocytes were still arrested in the metaphase-II stage 10 min after sacrificing the donor rats, but the proportion of oocytes progressing to anaphase-II or telophase-II increased in a time-dependent manner (Figure 10.1A) [8]. A significant decrease of the $p34^{cdc2}$ kinase activity was detected in the cultured rat oocytes; such a decrease in the cultured mouse oocytes occurred to a lesser extent (Figure 10.1B) [9]. The

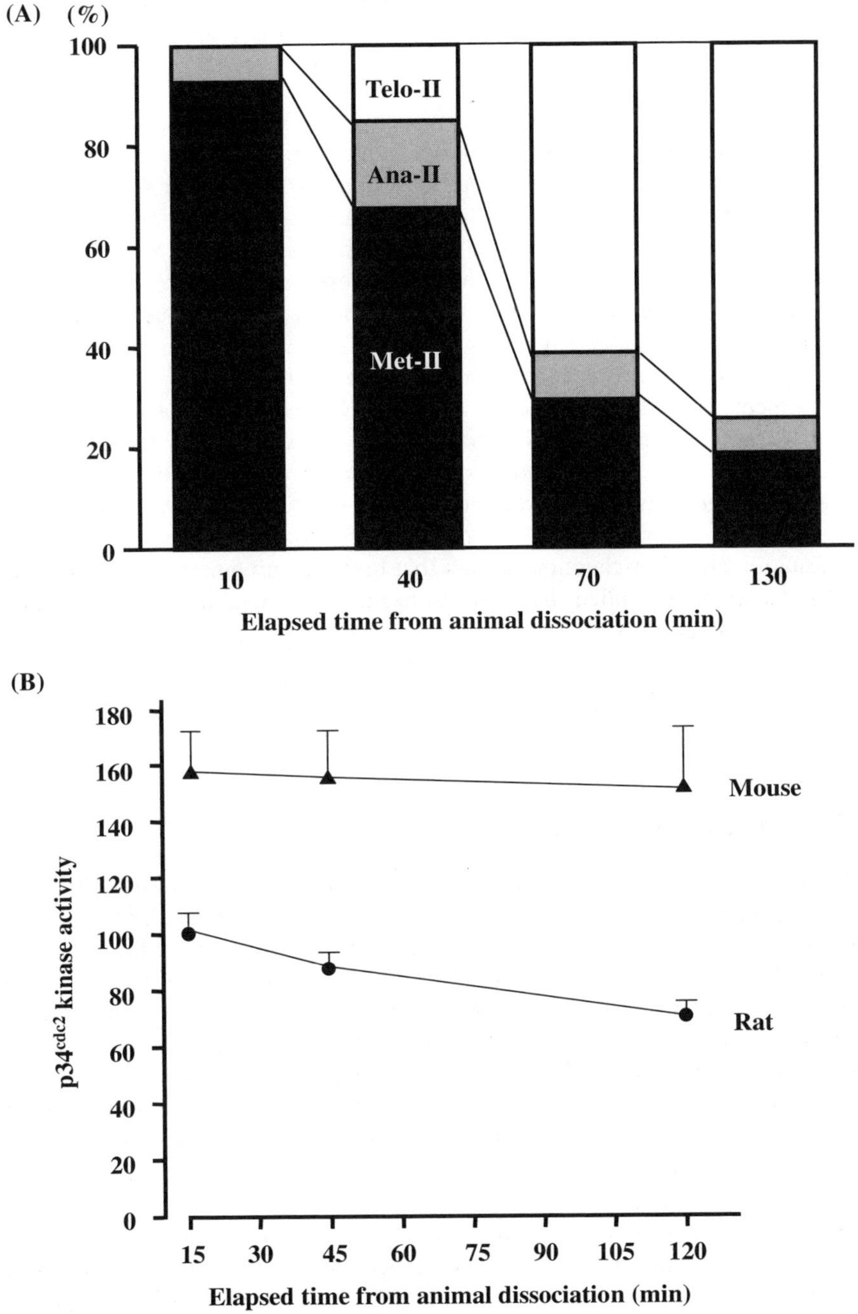

FIGURE 10.1 (A) Spontaneous activation of rat oocytes in culture (From Kato, M. et al., *J. Reprod. Dev.*, 47, 407, 2001) and (B) kinetics of p34^{cdc2} kinase activity of rat and mouse oocytes (From Ito, J. et al., *Reproduction*, 129, 171, 2005.)

relative difference in the maturation-promoting factor (MPF) baseline level between the rat and mouse oocytes may be especially important to note.

10.3.2 Induced Activation

Although spermatozoa carry the oocyte activating factors, somatic cell components are unable to activate oocytes following injection or fusion. Therefore, enucleated oocytes receiving adult somatic cells during the cloning process need to be activated artificially for their further development. Puromycin and chloral hydrate [7] and a Ca^{2+}/Mg^{2+}-free culture condition [10] have been reported to be potent activators that could trigger PN formation and first cleavage of the spontaneously activated rat oocytes. Strontium chloride (Sr^{2+}) has been used to activate enucleated rat oocytes receiving cumulus cell nuclei [8,11], as in mouse cloning [3].

Treatment with ethanol + cycloheximide (CHX) [12], Sr^{2+} + CHX [13], direct current pulses (DC) + 6-dimethylaminopurine (6-DMAP) [14,15], and ionomycin + 6-DMAP [16] resulted in the parthenogenetic development into blastocysts of the rat oocytes. Either activation treatment with Sr^{2+} or DC + 6-DMAP also supported the full-term development of the rat oocytes following round spermatid injection [17]. Thus, electrical or chemical stimuli that had been used to activate other mammalian oocytes were applied to rat oocytes with minor modifications.

10.4 NUCLEAR TRANSPLANTATION IN THE RAT

10.4.1 Cloned Rat Production Using Embryonic Cells

Research involving cloned rat production using zygotic/embryonic blastomeres has been very limited, probably due to the narrow applications of the resultant individuals. There are only two publications of successful NT in rats using nuclei from zygotic/embryonic cells. In 1988, Kono et al. [18] reported a high survival rate for reconstructed zygotes following electrofusion of zygotic karyoplasts into enucleated zygotes (60%, 126/210) and a high first cleavage rate of the NT zygotes *in vitro* (73%, 92/126), using Fischer (F334/DuCrj) rats. Nine live rat offspring were delivered after the transfer of 120 NT zygotes (7.5%). In the same study, the transfer of embryonic karyoplasts from blastomeres at various stages of development (two- to eight-cell embryos) into recipient zygote cytoplasms did not result in any offspring, suggesting that the developmental match between donor karyoplasts and recipient cytoplasms is important for the successful production of embryo-derived cloned rats.

Roh et al. [19] reported in the abstract form of the 2003 IETS meeting that the transfer of rat zygotes reconstructed using a two-cell stage embryo as the recipient cytoplasm resulted in the births of three offspring. Of those, two offspring were alive with low dense hair around their heads and one died 1 day after birth. The relationship of these abnormalities with NT is unknown. In this study, the NT zygotes were produced by the cell fusion method (2 DC pulses at 150 V/mm for 60 μsec) using karyoplasts and cytoplasms derived from two-cell stage embryos of Sprague–Dawley (SD) rats. Enucleated one-cell stage zygotes also supported the

TABLE 10.1
Numerous Efforts to Produce Cloned Rats by Somatic Cell Nuclear Transplantation

No. of Oocytes				
Treated	**Reconstructed**	**Transferred**	**Live Born**	**Ref.**
282	235 (83%)	224	0 (0%)	8
Not determined from the paper		35	0 (0%)	11
2766	653 (24%)	269	0 (0%)	12
429	178 (42%)	No development beyond two-cell stage		14
2042	478 (23%)	456	0 (0%)	23

full-term development following fusion with karyoplasts from one-cell zygotes. The proportions of NT rat zygotes developing into blastocysts in rat one-cell embryo culture medium (R1ECM) did not significantly differ between the karyoplast–cytoplasm origins (two-cell embryos: 34%, 34/99, and one-cell zygotes: 45%, 18/40).

10.4.2 Single NT Attempts with Somatic Cells

At least four independent groups have tried to produce cloned rats using the single NT procedure (Table 10.1):

- Kato et al. [8] attempted cloning rats according to the Honolulu method. The nuclei of cumulus cells were injected into enucleated SD rat oocytes using piezo-driven micropipettes; the reconstructed oocytes were then activated with 1.25 m*M* Sr^{2+} for 6 h. One implantation site but no live offspring at 21 days postcoitum (dpc) was observed in the nine surrogate mothers receiving a total of 224 presumptive cloned zygotes (including 50 cleaved embryos).
- Iannaccone et al. [11] reported that none of the 35 zygotes produced by the Honolulu method (rat strain for oocytes: SD; donor cell type" genetically modified fibroblasts; activation method: 2.5 m*M* Sr^{2+} for 1.5 to 2 h) developed to live offspring; of 86 zygotes, 21 exhibited signs of implantation at 13.5 dpc.
- Jiang et al. [14] obtained a total of 25 cleaved embryos from 178 zygotes reconstructed by injecting nuclei of ACI cell line cells or Wistar cumulus cells into enucleated Wistar oocytes and by activating with 2 DC (100 V/mm, 99 μsec) + 2 m*M* 6-DMAP for 2 to 4 h. These embryos did not progress to the later developmental stage in R1ECM.
- Hayes et al. [12] have conducted a series of large-scale experiments for cloning rats. As the best group, oocytes recovered from 4-week-old SD rats 12 h after human chorionic gonadotropin (hCG) injection were enucleated and cumulus cell nuclei were injected. After 30 to 60 min, the

reconstructed zygotes were activated with 8% ethanol for 5 min + 35 μ*M* CHX for 4 h. A total of 653 reconstructed zygotes survived enucleation, nuclear injection, and chemical activation out of subjected 2766 oocytes. The transfer of 269 zygotes, including 36 cleaved embryos, resulted in no live offspring. In a separate experiment, 492 oocytes were first activated and injected with cumulus cell nuclei and then enucleated. A total of 112 reconstructed zygotes, including 35 cleaved embryos, was produced, but transfer to surrogate mothers was not performed.

10.4.3 Serial NT Attempts with Somatic Cells

The keys to the successful production of cloned mice by the Honolulu method have been considered by the authors [3]; these include promoting the premature chromosome condensation (PCC) (Figure 10.2) and the chromosome formation of injected nuclei and the subsequent PN-like vesicle formation, as well as directly exposing injected nuclei to reprogramming factors present in the cytoplasm of the recipient oocytes. Therefore, the failure of NT rat embryos to develop into full-term offspring in the earlier studies may be due to microinjecting cell nuclei into oocytes that have already been spontaneously activated. The PCC would not occur even if oocytes with decreased MPF activity were enucleated and received somatic cell nuclei.

From this point of view, some possible factors influencing the PCC of cumulus cell nuclei injected into rat oocytes were examined (Table 10.2) [20]. The injected cumulus cell nucleus was distinguished from the oocyte nucleus by the position of the opening in the zona pellucida at the nuclear microinjection site. The potential of rat oocytes to support the PCC of injected nuclei depends on the characteristics of the oocytes, such as age or strain of the donor rats, as well as timing of the oocyte recovery. Based on these experiments, oocytes recovered 14 h after hCG injection

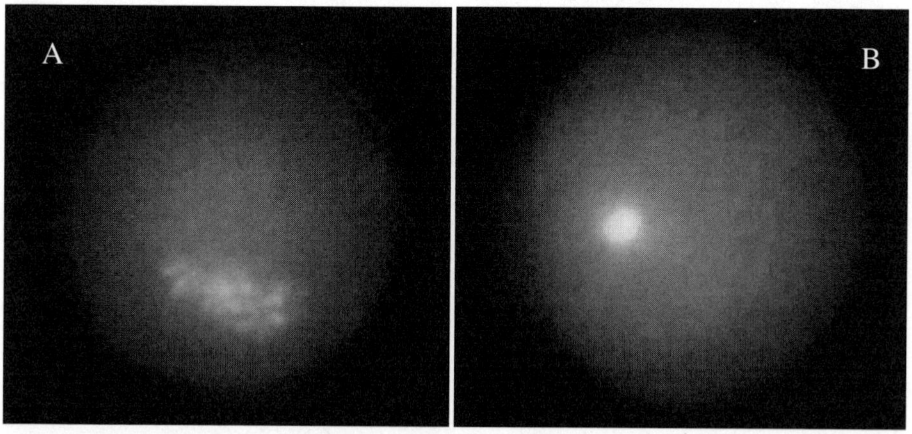

FIGURE 10.2 Morphological change in rat somatic cell nuclei injected into enucleated (A) mouse (PCC positive) and (B) rat (PCC negative) oocytes.

TABLE 10.2
Factors Affecting Chromosome Condensation of Donor Cell Nuclei in Rat Oocytes[a]

Factors	Outcomes with % of PCC-Positive Oocytes
Donor rat strains	Wistar (46%) vs. LEW (41%) vs. F344 (25%) vs. Donryu (17%)
Age of oocyte donors	4–5 weeks old (49%) vs. >10 weeks old (11%)
Timing of oocyte recovery	14 h post-hCG (44%) vs. 17 h post-hCG (7%)
Time until completing injection	15–45 min (43%) vs. 45–120 min (18%)
Inhibition of cyclin degradation	ALLN presence (48%) vs. ALLN absence (37%)
Enucleation (in case of rats)	Intact oocytes (44%) vs. enucleated oocytes (0%)
Enucleation (in case of mice)	Intact oocytes (97%) vs. enucleated oocytes (93%)

[a] Modified from Ito, J. et al., *Reproduction*, Nagasaki, 129, 171, 2005; Hirabayashi, M. et al., *J. Reprod. Dev.*, 49, 121, 2003; and Hirabayashi, M. et al., *Cloning Stem Cells*, 5, 35, 2003.

from young 4- to 5-week-old Wistar or Lewis (LEW/Crj) strain rats need to be used for nuclear injection as soon as possible.

The treatment of oocytes with a neutral cysteine protease inhibitor, *N*-acethylleucylleucylnorleucinal (ALLN), to maintain the meiotic arrest may be beneficial. However, the PCC did not occur when the cell nuclei were injected into enucleated rat oocytes. This phenomenon in the rat is in contrast to that in the mouse, in which more than 90% of the oocytes supported the PCC of injected nuclei regardless of the presence or absence of the recipient metaphase plate.

Kwon and Kono [21] first reported the serial NT procedure in the mouse. In their procedure, donor nuclei were first transferred into enucleated unfertilized oocytes and a PN-like vesicle was allowed to form. In the second transfer, karyoplasts containing the PN-like vesicle were transferred into a cytoplast derived from pronuclear-stage zygotes. It is suggested that the donor nuclei are allowed time for reprogramming during the first NT, and that the cytoplasm of the pronuclear zygotes supports further development of the second NT embryos. After the serial NT of the metaphase-arrested fetal fibroblast nuclei, the production of stillborn fetuses with severe abnormalities was reduced [22].

The authors have attempted to produce cloned rats by combining the serial NT procedure with PCC induction in somatic cell nuclei [23]. In the first series, the nuclei of cumulus cells (n = 494) or Sertoli cells (n = 574) from F1 hybrid (SD x Dark–Agouti) rats were injected into enucleated BDF1 mouse oocytes. The heterozygous NT oocytes were kept for 1 h and then activated with 10 mM Sr^{2+} for 6 h. From these reconstructed zygotes, karyoplasts containing PN-like vesicles were transferred into Wistar zygote-derived cytoplasts by a DC pulse (20 V/mm, 20 μsec). The transfer of a total of 340 serial NT zygotes into surrogate mothers, including 206 cleaved embryos, resulted in only seven implantation sites.

In the second series, the nuclei of rat cumulus cells (n = 576) or Sertoli cells (n = 398) were injected into 20 μg/ml ALLN-treated intact rat oocytes; the recipient metaphase plate was then aspirated under UV light from the NT oocytes in which the PCC of the injected nuclei was observed. After activation of the NT oocytes with 1.25 mM Sr^{2+} for 6 h, the karyoplasts were fused with zygote-derived cytoplasts. The transfer of a total of 115 serial NT zygotes, including 37 cleaved embryos, resulted in four implantation sites but no live offspring.

10.4.4 Successful Production of Rats Cloned with Fetal Fibroblasts

Very recently, Zhou et al. [5] published a paper entitled "Generation of Fertile Cloned Rats by Regulating Oocyte Activation" in the journal *Science*. Based on the online early release (published on September 25, 2003), the successful protocol can be summarized as shown in Figure 10.3. Oocytes were recovered from 3- to 4-week-old OFA-SD rats 14 h after hCG injection in M2 medium containing 5 μM MG132, a protease inhibitor that blocks the first meiotic metaphase-anaphase transition. Most oocytes treated with the MG132 reversibly stabilized their metaphase-II plates up to 3 h. Primary rat embryonic fibroblasts isolated from CD-SD 12.5 dpc fetuses were used as donor cells after 1 to 3 passages. Prior to the NT, the fibroblasts were treated with 0.05 μg/ml demecolcin for 2 h to synchronize the mitosis.

Within 30 min of MG132 removal, a fibroblast cell nucleus was injected into the oocytes on the opposite side of the recipient metaphase location. Before the injection pipette was completely withdrawn from the oocyte, the recipient metaphase plate was aspirated into the same pipette. This protocol made it possible to replace donor karyoplasts with oocyte chromosomes in one step. The NT oocytes were activated with 150 μM butyrolactone-I, a $p34^{cdc2}$ specific kinase inhibitor, for 2 h. Embryos cleaved in R1ECM were transferred to OFA-SD 0.5 dpc pseudopregnant females.

A total of 876 reconstructed zygotes were transferred to 12 surrogate mothers, resulting in 16 fetuses at 12.5 to 14.5 dpc. Thirteen of the fetuses were viable, with heart beating. In the next series, of 129 zygotes that were allowed to go to term in two surrogate mothers, three (2.3%) live male offspring were delivered. One pup died a few hours after birth, but his two brothers survived to sexual maturity and produced progeny. In the last series, two female cloned rats resulted from the transfer of c.a. 100 zygotes to one surrogate mother. They also produced progeny. Thus, the authors demonstrated that the single successive manipulation of injection and enucleation combined with MG132/butyrolactone-I treatment was effective for the production of fertile cloned rats of both sexes.

10.5 FUTURE PERSPECTIVE OF RAT CLONING

It is undoubtedly true that cloned rat individuals are beneficial experimental materials for fundamental and clinical studies in the medical field. What is the most useful application of cloning technology in this species? The answer would be to contribute in the production of "knock-out (KO)" animals by combining with homologous recombination-based gene targeting technology in cultured cells. Because the rat is used, especially in the research field of neuroscience, developments in the methodology to produce KO rats have been long desired.

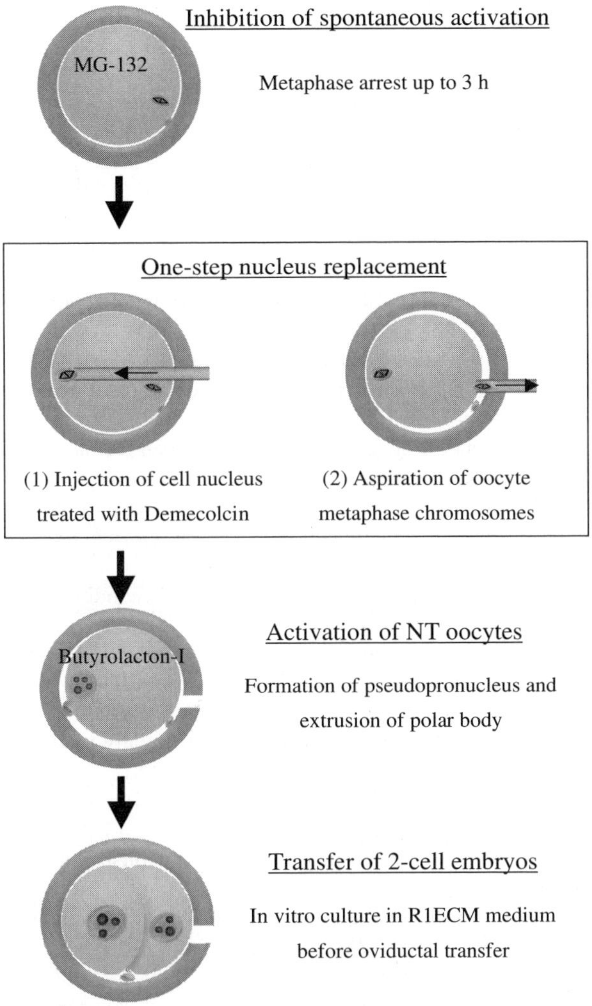

FIGURE 10.3 A successful protocol to produce cloned rats. (Reported by Zhou, Q. et al., *Science*, 302, 1179, 2003.)

The most convenient method for the production of KO animals is to use gene-targeted embryonic stem (ES) cells in germline chimera production; however, ES cell lines have not yet been established in the rat. The male germline stem cell line, capable of culturing and differentiating *in vitro*, is an alternative source to produce KO animals (via microinsemination), but such studies have been limited in the mouse [24–26]. Therefore, reproducibility of Zhou and colleagues' NT protocol to clone rats would have a strong impact on the future progress of medical research and laboratory animal science. Understanding the fundamental aspects of cell cycle regulation during genome reprogramming in rat oocytes must be useful for improving the protocol for the production of somatic cell-derived cloned rats.

REFERENCES

1. Wilmut, I. et al., Viable offspring derived from fetal and adult mammalian cells, *Nature*, 385, 810, 1997.
2. Kato, Y. et al., Eight calves cloned from somatic cells of a single adult, *Science*, 282, 2095, 1998.
3. Wakayama, T. et al., Full-term development of mice from enucleated oocytes injected with cumulus cell nuclei, *Nature*, 394, 369, 1998.
4. Onishi, A. et al., Pig cloning by microinjection of fetal fibroblast nuclei, *Science*, 289, 1188, 2000.
5. Zhou, Q. et al., Generation of fertile cloned rats by regulating oocyte activation, *Science*, 302, 1179, 2003.
6. Gill III, T.J. et al., The rat as an experimental animal, *Science*, 245, 269, 1989.
7. Zernicka–Goetz, M., Spontaneous and induced activation of rat oocytes, *Mol. Reprod. Dev.*, 28, 169, 1991.
8. Kato, M. et al., Strontium-induced activation regimen for rat oocytes in somatic cell nuclear transplantation, *J. Reprod. Dev.*, 47, 407, 2001.
9. Ito, J. et al., Contribution of high $p34^{cdc2}$ kinase activity to premature chromosome condensation of injected somatic cell nuclei in rat oocytes, *Reproduction*, 129, 171, 2005.
10. Takeuchi, A. et al., Effect of Ca^{2+}/Mg^{2+}-free culture condition on spontaneous first cleavage of rat oocytes, *J. Reprod. Dev.*, 48, 243, 2002.
11. Iannaccone, P., Taborn, G., and Garton, R., Preimplantation and postimplantation development of rat embryos cloned with cumulus cells and fibroblasts, *Zygote*, 9, 135, 2001.
12. Hayes, E. et al., Nuclear transfer of adult and genetically modified fetal cells of the rat, *Physiol. Genomics*, 5, 193, 2001.
13. Roh, S. et al., Parthenogenetic activation of rat oocytes and their development *in vitro*, *Reprod. Fertil. Develop.*, 15, 135, 2003.
14. Jiang, J.Y. et al., Nuclear transfer in rats using an established embryonic cell line and cumulus cells, *J. Reprod. Dev.*, 48, 505, 2002.
15. Jiang, J.Y. et al., Parthenogenetic activation and subsequent development of rat oocytes *in vitro*, *Mol. Reprod. Dev.*, 61, 120, 2002.
16. Mizutani, E. et al., Determination of optimal conditions for parthenogenetic activation and subsequent development of rat oocytes *in vitro*, *J. Reprod. Dev.*, 50, 139, 2004.
17. Kato, M. et al., Effect of activation regimens for rat oocytes on full-term development after round spermatid injection, *Contem. Top. Lab. Anim. Sci.*, 43/2, 13, 2004.
18. Kono, T., Shioda, Y., and Tsunoda, Y., Nuclear transplantation of rat embryos, *J. Exp. Zool.*, 248, 303, 1988.
19. Roh, S. et al., Birth of rats by nuclear transplantation using two-cell stage embryos as donor nucleus and recipient cytoplasm, *Theriogenology*, 59, 283, 2003.
20. Hirabayashi, M. et al., Factors affecting premature chromosome condensation of cumulus cell nuclei injected into rat oocytes, *J. Reprod. Dev.*, 49, 121, 2003.
21. Kwon, O. and Kono, T., Production of live young by serial nuclear transfer with mitotic stage of donor nuclei in mice, *Proc. Natl. Acad. Sci. U.S.A.*, 93, 13010, 1996.
22. Ono, Y. et al., Cloned mice from fetal fibroblast cells arrested at metaphase by a serial nuclear transfer, *Biol. Reprod.*, 64, 44, 2001.
23. Hirabayashi, M. et al., Factors influencing chromosome condensation and development of cloned rat embryos, *Cloning Stem Cells*, 5, 35, 2003.

24. Feng, L.X. et al., Generation and *in vitro* differentiation of a spermatigonial cell line, *Science*, 297, 392, 2002.
25. Toyooka, Y. et al., Embryonic stem cells can form germ cells *in vitro*, *Proc. Natl. Acad. Sci. U.S.A.*, 100, 11457, 2003.
26. Geijsen, N. et al., Derivation of embryonic germ cells and male gametes from embryonic stem cells, *Nature*, 427, 148, 2004.

Index

C

D

E

F

G

K

L

M

N

O

P

R

S